Electromagnetic Band Gap Structures in Antenna Engineering

This comprehensive, applications-oriented survey of the state of the art in Electromagnetic Band Gap (EBG) engineering explains the theory, analysis, and design of EBG structures. It helps you to understand EBG applications in antenna engineering through an abundance of novel antenna concepts, a wealth of practical examples, and complete design details. You discover a customized finite difference time domain (FDTD) method of EBG analysis, for which accurate and efficient electromagnetic software is supplied (www.cambridge.org/yang) to provide a powerful computational engine for your EBG desgins. The first book covering EBG structures and their antenna applications, this provides a dynamic resource for engineers, and researchers and graduate students working in antennas, electromagnetics and microwaves.

Fan Yang is Assistant Professor of the Electrical Engineering Department at the University of Mississippi. Dr. Yang received Young Scientist Awards in the 2005 General Assembly of International Union of Radio Science (URSI) and in the 2007 International Symposium on Electromagnetic Theory.

Yahya Rahmat-Samii is a Distinguished Professor, holder of the Northrop-Grumman Chair in Electromagnetics, and past Chairman of the Electrical Engineering Department at the University of California, Los Angeles (UCLA). He has received numerous recognitions and awards including IEEE Fellow in 1985, IEEE Third Millennium Medal, the 2005 URSI Booker Gold Medal, the 2007 Chen-To Tai Distinguished Educator Award of the IEEE Antennas and Propagation Society, and membership of the US National Academy of Engineering.

The Cambridge RF and Microwave Engineering Series

Series Editor,
Steve C. Cripps, Hywave Associates

Peter Aaen, Jaime A. Plá, and John Wood, *Modeling and Characterization of RF and Microwave Power FETs*
Enrico Rubiola, *Phase Noise and Frequency Stability in Oscillators*
Dominique Schreurs, Máirtín O'Droma, Anthony A. Goacher, and Michael Gadringer, *RF Amplifier Behavioral Modeling*
Fan Yang and Yahya Rahmat-Samii, *Electromagnetic Band Gap Structures in Antenna Engineering*

Forthcoming
Sorin Voinigescu and Timothy Dickson, *High-Frequency Integrated Circuits*
Debabani Choudhury, *Millimeter Waves for Commercial Applications*
J. Stephenson Kenney, *RF Power Amplifier Design and Linearization*
David B. Leeson, *Microwave Systems and Engineering*
Stepan Lucyszyn, *Advanced RF MEMS*
Earl McCune, *Practical Digital Wireless Communications Signals*
Allen Podell and Sudipto Chakraborty, *Practical Radio Design Techniques*
Patrick Roblin, *Nonlinear RF Circuits and the Large-Signal Network Analyzer*
Dominique Schreurs, *Microwave Techniques for Microelectronics*
John L. B. Walker, *Handbook of RF and Microwave Solid-State Power Amplifiers*

Electromagnetic Band Gap Structures in Antenna Engineering

FAN YANG
University of Mississippi

YAHYA RAHMAT-SAMII
University of California at Los Angeles

CAMBRIDGE UNIVERSITY PRESS
Cambridge, New York, Melbourne, Madrid, Cape Town,
Singapore, São Paulo, Delhi, Mexico City

Cambridge University Press
The Edinburgh Building, Cambridge CB2 8RU, UK

Published in the United States of America by Cambridge University Press, New York

www.cambridge.org
Information on this title: www.cambridge.org/9780521889919

© Cambridge University Press 2009

This publication is in copyright. Subject to statutory exception
and to the provisions of relevant collective licensing agreements,
no reproduction of any part may take place without the written
permission of Cambridge University Press.

First published 2009

A catalogue record for this publication is available from the British Library

Library of Congress Cataloguing in Publication Data
Yang, Fan, 1975–
Electromagnetic band gap structures in antenna engineering / Fan Yang, Yahya Rahmat-Samii.
    p. cm.
Includes index.
ISBN 978-0-521-88991-9 (hbk. : alk. paper)   1. Antennas (Electronics) – Design and construction.
2. Wide gap semiconductors.   I. Rahmat-Samii, Yahya.   II. Title.
TK7871.6.Y35   2008
621.382′4–dc22   2008031949

ISBN 978-0-521-88991-9 Hardback

Cambridge University Press has no responsibility for the persistence or
accuracy of URLs for external or third-party internet websites referred to in
this publication, and does not guarantee that any content on such websites is,
or will remain, accurate or appropriate. Information regarding prices, travel
timetables, and other factual information given in this work is correct at
the time of first printing but Cambridge University Press does not guarantee
the accuracy of such information thereafter.

Contents

Preface		*page* ix
Acknowledgements		xi
Abbreviations		xii

1	**Introduction**	1
1.1	Background	1
1.2	Electromagnetic band gap (EBG) structures	2
	1.2.1 EBG definition	2
	1.2.2 EBG and metamaterials	4
1.3	Analysis methods for EBG structures	6
1.4	EBG applications in antenna engineering	8
	1.4.1 Antenna substrates for surface wave suppressions	8
	1.4.2 Antenna substrates for efficient low profile wire antenna designs	9
	1.4.3 Reflection/transmission surfaces for high gain antennas	10

2	**FDTD method for periodic structure analysis**	14
2.1	FDTD fundamentals	14
	2.1.1 Introduction	14
	2.1.2 Yee's cell and updating scheme	15
	2.1.3 Absorbing boundary conditions: PML	18
	2.1.4 FDTD excitation	22
	2.1.5 Extraction of characteristic parameters	23
2.2	Periodic boundary conditions	24
	2.2.1 Fundamental challenges in PBC	24
	2.2.2 Overview of various PBCs	25
	2.2.3 Constant k_x method for scattering analysis	26
2.3	Guided wave analysis	30
	2.3.1 Problem statement	30
	2.3.2 Brillouin zone for periodic waveguides	31
	2.3.3 Examples	33
2.4	Plane wave scattering analysis	37
	2.4.1 Problem statement	38
	2.4.2 Plane wave excitation	39
	2.4.3 Examples	41

	2.5	A unified approach: hybrid FDTD/ARMA method	45
		2.5.1 A unified approach for guided wave and scattering analysis	45
		2.5.2 ARMA estimator	49
		2.5.3 Examples	51
	2.6	Projects	54

3 EBG characterizations and classifications 59

	3.1	Resonant circuit models for EBG structures	59
		3.1.1 Effective medium model with lumped LC elements	59
		3.1.2 Transmission line model for surface waves	61
		3.1.3 Transmission line model for plane waves	62
	3.2	Graphic representation of frequency band gap	63
		3.2.1 FDTD model	63
		3.2.2 Near field distributions inside and outside the frequency band gap	65
	3.3	Frequency band gap for surface wave propagation	67
		3.3.1 Dispersion diagram	67
		3.3.2 Surface wave band gap	68
	3.4	In-phase reflection for plane wave incidence	69
		3.4.1 Reflection phase	69
		3.4.2 EBG reflection phase: normal incidence	70
		3.4.3 EBG reflection phase: oblique incidence	71
	3.5	Soft and hard surfaces	74
		3.5.1 Impedance and reflection coefficient of a periodic ground plane	75
		3.5.2 Soft and hard operations	77
		3.5.3 Examples	80
	3.6	Classifications of various EBG structures	84
	3.7	Project	85

4 Designs and optimizations of EBG structures 87

	4.1	Parametric study of a mushroom-like EBG structure	87
		4.1.1 Patch width effect	87
		4.1.2 Gap width effect	89
		4.1.3 Substrate thickness effect	89
		4.1.4 Substrate permittivity effect	90
	4.2	Comparison of mushroom and uni-planar EBG designs	91
	4.3	Polarization-dependent EBG surface designs	95
		4.3.1 Rectangular patch EBG surface	95
		4.3.2 Slot loaded EBG surface	97
		4.3.3 EBG surface with offset vias	97
		4.3.4 An example application: PDEBG reflector	99
	4.4	Compact spiral EBG designs	103
		4.4.1 Single spiral design	103

		4.4.2 Double spiral design	105
		4.4.3 Four-arm spiral design	105
	4.5	Dual layer EBG designs	107
	4.6	Particle swarm optimization (PSO) of EBG structures	112
		4.6.1 Particle swarm optimization: a framework	112
		4.6.2 Optimization for a desired frequency with a $+90°$ reflection phase	113
		4.6.3 Optimization for a miniaturized EBG structure	117
		4.6.4 General steps of EBG optimization problems using PSO	118
	4.7	Advanced EBG surface designs	120
		4.7.1 Space filling curve EBG designs	120
		4.7.2 Multi-band EBG surface designs	120
		4.7.3 Tunable EBG surface designs	120
	4.8	Projects	124

5 Patch antennas with EBG structures — 127

	5.1	Patch antennas on high permittivity substrate	127
	5.2	Gain enhancement of a single patch antenna	130
		5.2.1 Patch antenna surrounded by EBG structures	130
		5.2.2 Circularly polarized patch antenna design	132
		5.2.3 Various EBG patch antenna designs	136
	5.3	Mutual coupling reduction of a patch array	138
		5.3.1 Mutual coupling between patch antennas on high dielectric constant substrate	139
		5.3.2 Mutual coupling reduction by the EBG structure	142
		5.3.3 More design examples	147
	5.4	EBG patch antenna applications	149
		5.4.1 EBG patch antenna for high precision GPS applications	149
		5.4.2 EBG patch antenna for wearable electronics	149
		5.4.3 EBG patch antennas in phased arrays for scan blindness elimination	151
	5.5	Projects	153

6 Low profile wire antennas on EBG ground plane — 156

	6.1	Dipole antenna on EBG ground plane	156
		6.1.1 Comparison of PEC, PMC, and EBG ground planes	156
		6.1.2 Operational bandwidth selection	158
		6.1.3 Parametric studies	161
	6.2	Low profile antennas: wire-EBG antenna vs. patch antenna	164
		6.2.1 Two types of low profile antennas	164
		6.2.2 Performance comparison between wire-EBG and patch antennas	166
		6.2.3 A dual band wire-EBG antenna design	169

6.3	Circularly polarized curl antenna on EBG ground plane		171
	6.3.1	Performance of curl antennas over PEC and EBG ground planes	172
	6.3.2	Parametric studies of curl antennas over the EBG surface	175
	6.3.3	Experimental demonstration	178
6.4	Dipole antenna on a PDEBG ground plane for circular polarization		180
	6.4.1	Radiation mechanism of CP dipole antenna	181
	6.4.2	Experimental results	182
6.5	Reconfigurable bent monopole with radiation pattern diversity		185
	6.5.1	Bent monopole antenna on EBG ground plane	186
	6.5.2	Reconfigurable design for one-dimensional beam switch	188
	6.5.3	Reconfigurable design for two-dimensional beam switch	191
6.6	Printed dipole antenna with a semi-EBG ground plane		191
	6.6.1	Dipole antenna near the edge of a PEC ground plane	193
	6.6.2	Enhanced performance of dipole antenna near the edge of an EBG ground plane	194
	6.6.3	Printed dipole antenna with a semi-EBG ground plane	195
6.7	Summary		200
6.8	Projects		200

7 Surface wave antennas — 203

7.1	A grounded slab loaded with periodic patches		203
	7.1.1	Comparison of two artificial ground planes	203
	7.1.2	Surface waves in the grounded slab with periodic patch loading	206
7.2	Dipole-fed surface wave antennas		209
	7.2.1	Performance of a low profile dipole on a patch-loaded grounded slab	209
	7.2.2	Radiation mechanism: the surface wave antenna	212
	7.2.3	Effect of the finite artificial ground plane	215
	7.2.4	Comparison between the surface wave antenna and vertical monopole antenna	217
7.3	Patch-fed surface wave antennas		217
	7.3.1	Comparison between a circular microstrip antenna and a patch-fed SWA	218
	7.3.2	Experimental demonstration	223
7.4	Dual band surface wave antenna		223
	7.4.1	Crosspatch-fed surface wave antenna	226
	7.4.2	Modified crosspatch-fed surface wave antenna for dual band operation	228
7.5	Projects		236

Appendix: EBG literature review — 238
Index — 261

Preface

In recent years, electromagnetic band gap (EBG) structures have attracted increasing interests because of their desirable electromagnetic properties that cannot be observed in natural materials. In this respect, EBG structures are a subset of metamaterials. Diverse research activities on EBG structures are on the rise in the electromagnetics and antenna community, and a wide range of applications have been reported, such as low profile antennas, active phased arrays, TEM waveguides, and microwave filters. We believe that the time is right for a focused book reviewing the state of the art on electromagnetic band gap (EBG) structures and their important applications in antenna engineering.

The goal of this book is to provide scientists and engineers with an up-to-date knowledge on the theories, analyses, and applications of EBG structures. Specifically, this book will cover the following topics:

- a detailed overview of the EBG research history and important results;
- an advanced presentation on the unique features of EBG structures;
- an accurate and efficient numerical algorithm for EBG analysis and an evolutionary optimization technique for EBG design;
- a wealth of examples illustrating potential applications of EBG structures in antenna engineering.

The book is organized into seven chapters and one appendix. Chapter 1 introduces the background and basic properties of EBG structures. The EBG analysis methods and antenna applications are also summarized.

In Chapter 2, the finite difference time domain (FDTD) method is presented with a focus on periodic boundary conditions (PBC), which is used as an efficient computation engine for the analysis of periodic structures. The fundamentals of the FDTD method are reviewed and a constant k_x (spectral) method is discussed to model the PBC. A hybrid FDTD/ARMA scheme is introduced to unify the guided wave and plane wave analysis and improve the simulation efficiency.

Chapter 3 illustrates some interesting properties of EBG structures. The band gap features are clearly visualized from the near field distributions. The dispersion diagram and reflection phase for both normal and oblique incidences are presented. The soft and hard properties of the EBG ground plane are also discussed. A classification of various EBG structures is provided at the end of the chapter.

Chapter 4 presents how to achieve the desired characteristics by properly designing the EBG structures. A parametric study on the mushroom-like EBG structure is performed first, followed by a comparison between two popularly used planar EBG structures, mushroom-like EBG surface and uni-planar EBG surface. Novel EBG designs such as polarization-dependent EBG (PDEBG), compact spiral EBG, and stacked EBG structures are also studied. Furthermore, utilizations of the particle swarm optimization (PSO) technique are demonstrated in EBG synthesis.

The applications of EBG structures in antenna engineering are presented in Chapters 5, 6 and 7. In Chapter 5, the EBG structures are integrated into microstrip patch antenna designs. The surface wave band gap property of EBG helps to increase the antenna gain, minimize the back lobe, and reduce mutual coupling. Some applications of EBG patch antenna designs in high precision GPS receivers, wearable electronics, and phased array systems are highlighted at the end of the chapter.

Chapter 6 introduces a novel type of antennas: low profile wire antennas on an EBG ground plane. Using the in-phase reflection feature of the EBG structure, the radiation efficiency of wire antennas near a ground plane can be greatly improved. A series of design examples are illustrated, including dipole, monopole, and curl. Various functionalities have been realized, such as dual band operation, circular polarization, and pattern diversity.

Chapter 7 presents a grounded slab loaded with periodic patches that can enhance the surface waves along a thin ground plane. Using this property, a low profile surface wave antenna (SWA) is designed, which achieves a monopole-like radiation pattern with a null in the broadside direction. Different feed techniques are explored and a dual band SWA is developed.

In the Appendix, a comprehensive literature review is presented based on nearly 300 references. This is to help both the seasoned and new comers in this research arena to establish a clear picture of the EBG developments and identify published work related to their own research interests. We regret if we have missed some of the publications as it has been very hard to identify all the research and development works that have been conducted in various international organizations.

We hope that the readers find this book useful and we welcome all their constructive suggestions.

F. Yang and Y. Rahmat-Samii

Acknowledgements

One of the authors (Rahmat-Samii) would like to express his sincere gratitude to his former students at the UCLA Antenna Research and Analysis Laboratory whose research contributions under his supervision were the basis of this book. In particular, special appreciation is extended to Michael Jensen, Joseph Colburn, Alon Barlevy, Zhan Li, Fan Yang, Hossein Mosallaei, Amir Aminian, and Nanbo Jin. Nanbo Jin is particularly thanked for his authoring Section 4.6, "Particle swarm optimization (PSO) of EBG structures" of this book.

The other author (Yang) also thanks his colleagues for the helpful discussions and productive collaborations in this research, including Ji Chen, from the University of Houston, Atef Elsherbeni, Ahmed Kishk, Veysel Demir, Yanghyo Kim, and Asem Al-Zoubi from the University of Mississippi. In particular, Yanghyo Kim's contribution on the graphic user interface (GUI) and manual of the periodic FDTD software is greatly appreciated.

We also thank the Cambridge University Press staff, Julie Lancashire, and Sabine Koch, for their continuous support during the compilation of this book.

Abbreviations

ABC	Absorbing Boundary Condition
ANN	Artificial Neural Network
AMC	Artificial Magnetic Conductor
AR	Axial Ratio
ARMA	Auto-Regressive Moving Average
CP	Circular Polarization
DNG	Double NeGative
DBSWA	Dual Band Surface Wave Antenna
EBG	Electromagnetic Band Gap
EMI	ElectroMagnetic Interference
FDTD	Finite Difference Time Domain
FEM	Finite Element Method
FSS	Frequency Selective Surface
GA	Genetic Algorithm
GPS	Global Positioning System
GUI	Graphic User Interface
HIS	High Impedance Surface
LH	Left Handed
LHCP	Left Hand Circular Polarization
LTCC	Low Temperature Co-fired Ceramic
MEMS	Micro-Electro-Mechanical System
MMIC	Monolithic Microwave Integrated Circuit
MoM	Method of Moment
NRI	Negative Refractive Index
PBC	Periodic Boundary Condition
PBG	Photonic Band Gap
PCB	Printed Circuit Board
PDEBG	Polarization-Dependent Electromagnetic Band Gap
PEC	Perfect Electric Conductor
PMC	Perfect Magnetic Conductor
PFSWA	Patch-Fed Surface Wave Antenna
PML	Perfectly Matched Layers
PSO	Particle Swarm Optimization
RCS	Radar Cross Section

RHCP	Right Hand Circular Polarization
RFID	Radio Frequency IDentification
SFDTD	Spectral Finite Difference Time Domain
SWA	Surface Wave Antenna
TE	Transverse Electric
TEM	Transverse ElectroMagnetic
TM	Transverse Magnetic
WLAN	Wireless Local Area Network

1 Introduction

1.1 Background

Antenna designs have experienced enormous advances in the past several decades and they are still undergoing monumental developments. Many new technologies have emerged in the modern antenna design arena and one exciting breakthrough is the discovery/development of electromagnetic band gap (EBG) structures. The applications of EBG structures in antenna designs have become a thrilling topic for antenna scientists and engineers. This is the central focus of this book.

The recent explosion in antenna developments has been fueled by the increasing popularity of wireless communication systems and devices. From the traditional radio and TV broadcast systems to the advanced satellite system and wireless local area networks, wireless communications have evolved into an indispensable part of people's daily lives. Antennas play a paramount role in the development of modern wireless communication devices, ranging from cell phones to portable GPS navigators, and from the network cards of laptops to the receivers of satellite TVs. A series of design requirements, such as low profile, compact size, broad bandwidth, and multiple functionalities, keep on challenging antenna researchers and propelling the development of new antennas.

Progress in computational electromagnetics, as another important driving force, has substantially contributed to the rapid development of novel antenna designs. It has greatly expanded the antenna researchers' capabilities in improving and optimizing their designs efficiently. Various numerical techniques, such as the method of moments (MoM), finite element method (FEM), and the finite difference time domain (FDTD) method, have been well developed over the years. As a consequence, numerous commercial software packages have emerged. Nowadays with powerful personal computers and advanced numerical techniques or commercial software, antenna researchers are able to exploit complex engineered electromagnetic materials in antenna designs, resulting in many novel and efficient antenna structures.

For the same reasons, electromagnetic band gap (EBG) structures and their applications in antennas have become a new research direction in the antenna community. It was first proposed to respond to some antenna challenges in wireless communications. For example,

- How to suppress surface waves in the antenna ground plane?
- How to design an efficient low profile wire antenna near a ground plane?

- How to achieve a uniform field distribution in a rectangular waveguide?
- How to increase the gain of an antenna?

Discovery of EBG structures has revealed promising solutions to the above problems. Due to the complexity of the EBG structures, it is usually difficult to characterize them through analytical methods. Instead, full wave simulators that are based on advanced numerical methods have been popularly used in EBG analysis. Dispersion diagram, surface impedance, and reflection phase features are explored for different EBG structures. The interaction of antennas and EBG structures are extensively investigated. In summary, the EBG research has flourished since the beginning of this new millennium.

1.2 Electromagnetic band gap (EBG) structures

So, what are electromagnetic band gap (EBG) structures? This section addresses this question from two aspects: definition of EBG structures and the relation between EBG and metamaterials.

1.2.1 EBG definition

Periodic structures are abundant in nature, which have fascinated artists and scientists alike. When they interact with electromagnetic waves, exciting phenomena appear and amazing features result. In particular, characteristics such as frequency stop bands, pass bands, and band gaps could be identified. Reviewing the literature, one observes that various terminologies have been used depending on the domain of the applications. These applications are seen in filter designs, gratings, frequency selective surfaces (FSS) [1], photonic crystals [2] and photonic band gaps (PBG) [3], etc. We classify them under the broad terminology of *"Electromagnetic Band Gap (EBG)"* structures [4].

Generally speaking, electromagnetic band gap structures are defined as *artificial periodic (or sometimes non-periodic) objects that prevent/assist the propagation of electromagnetic waves in a specified band of frequency for all incident angles and all polarization states.*

EBG structures are usually realized by periodic arrangement of dielectric materials and metallic conductors. In general, they can be categorized into three groups according to their geometric configuration: (1) three-dimensional volumetric structures, (2) two-dimensional planar surfaces, and (3) one-dimensional transmission lines. Figure 1.1 shows two representative 3-D EBG structures: a woodpile structure consisting of square dielectric bars [5] and a multi-layer metallic tripod array [6]. Examples of 2-D EBG surfaces are plotted in Fig. 1.2: a mushroom-like surface [7] and a uni-planar design without vertical vias [8]. Figure 1.3 shows the one-dimensional EBG transmission line designs [9–10]. This book focuses more on the 2-D EBG surfaces, which have the advantages of low profile, light weight, and low fabrication cost, and are widely considered in antenna engineering.

The planar electromagnetic band gap (EBG) surfaces exhibit distinctive electromagnetic properties with respect to incident electromagnetic waves:

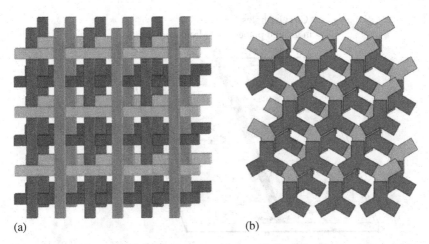

Fig. 1.1 Three-dimensional EBG structures: (a) a woodpile dielectric structure and (b) a multi-layer metallic tripod array.

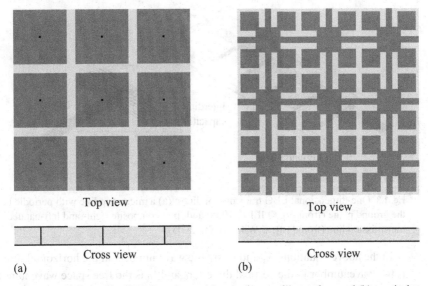

Fig. 1.2 Two-dimensional EBG surfaces: (a) a mushroom-like surface and (b) a uni-planar surface.

(1) When the incident wave is a surface wave ($k_x^2 + k_y^2 \leq k_0^2$, k_z is purely imaginary), the EBG structures show a frequency band gap through which the surface wave cannot propagate for any incident angles and polarization states. A typical dispersion diagram is shown in Fig. 1.4a.

(2) When the incident wave is a plane wave ($k_x^2 + k_y^2 \leq k_0^2$, k_z has a real value), the reflection phase of the EBG structures varies with frequency, as shown in Fig. 1.4b. At a certain frequency the reflection phase is zero degrees, which resembles a perfect magnetic conductor that does not exist in nature.

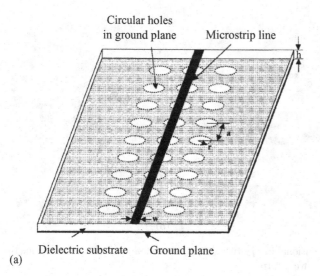

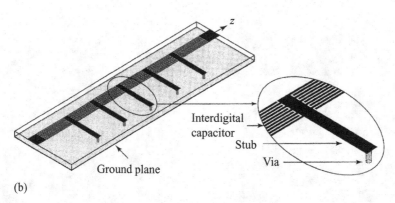

Fig. 1.3 One-dimensional EBG transmission lines: (a) a microstrip line with periodic holes on the ground plane (from [9], © IEEE, 1998) and (b) a composite right- and left-handed transmission line (from [10], © Wiley-IEEE, 2005).

In the above equations, k_x and k_y are the wavenumbers in the horizontal directions, k_z is the wavenumber in the vertical direction, and k_0 is the free space wavenumber.

1.2.2 EBG and metamaterials

Almost at the same time, another terminology, "*metamaterials*," also appeared and has become popular in the electromagnetics community [10–14]. The ancient Greek prefix, *meta* (meaning "beyond"), has been used to describe composite materials with unique features not readily available in nature. Depending on the exhibited electromagnetic properties, various names have been introduced in the literature, including:

- *Double negative (DNG) materials* with both negative permittivity and permeability;
- *Left-handed (LH) materials* inside which the electric field direction, magnetic field direction, and propagation direction satisfy a left-hand relation;

1.2 Electromagnetic band gap (EBG) structures

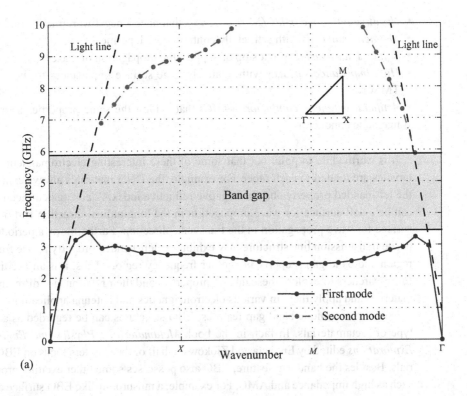

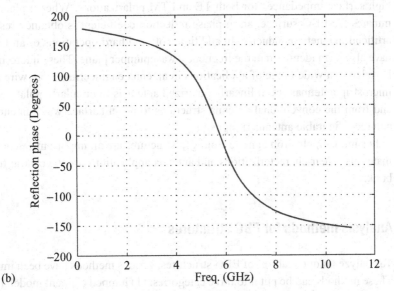

Fig. 1.4 A planar EBG surface exhibits (a) a surface wave band gap and (b) an in-phase reflection coefficient for plane wave incidence.

- *Negative refractive index (NRI) materials* that have a negative refractive index;
- *Magneto materials* with artificially controlled high permeability;
- *Soft and hard surfaces* that stop or support the propagation of waves;
- *High impedance surfaces* with relatively large surface impedances for both TE and TM waves;
- *Artificial magnetic conductors (AMC)* that exhibit the same properties as a perfect magnetic conductor.

It is worthwhile to point out that some of these interesting electromagnetic characteristics are related to each other. For example, the DNG materials always exhibit both the left-handed property and the negative refractive index. A corrugated metal surface can be a soft surface for wave propagation in the longitudinal direction and be a hard surface for wave propagation in the transverse direction. Furthermore, a periodic composite transmission line structure may exhibit the left-handed property in one frequency region and band gap property in another frequency region. Thus, it is an exciting area for researchers to explore these unique properties and their relations for different metamaterials and apply them in various electromagnetics and antenna applications.

Due to their unique band gap features, EBG structures can be regarded as a special type of metamaterials. In fact, in the book *Metamaterials: Physics and Engineering Explorations* edited by Engheta and Ziokowski, half of the chapters focus on EBG materials. Besides the band gap feature, EBG also possesses some other exciting properties, such as high impedance and AMC. For example, a mushroom-like EBG surface exhibits high surface impedances for both TE and TM polarizations. When a plane wave illuminates the EBG surface, an in-phase reflection coefficient is obtained resembling an artificial magnetic conductor. In addition, soft and hard operations of an EBG surface have also been identified in the frequency-wavenumber plane. These interesting features have led to a wide range of applications in antenna engineering, from wire antennas to microstrip antennas, from linearly polarized antennas to circularly polarized antennas, and from the conventional antenna structures to novel surface wave antenna concepts and reconfigurable antenna designs.

In summary, electromagnetic band gap structures are an important category of metamaterials. Their characterizations and antenna applications are the central focus of this book.

1.3 Analysis methods for EBG structures

To analyze unique features of EBG structures, various methods have been implemented. These methods can be put into three categories: (1) lumped element model, (2) periodic transmission line method, and (3) full wave numerical methods. The lumped element model is the simplest one that describes the EBG structure as an LC resonant circuit [15], as shown in Fig. 1.5. The values of the inductance L and capacitance C are determined by the EBG geometry and its resonance behavior is used to explain the band gap feature

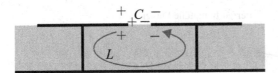

Fig. 1.5 Lumped LC model for EBG analysis.

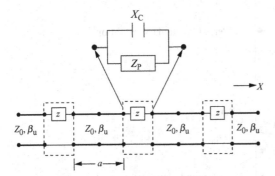

Fig. 1.6 Periodic transmission line method for EBG analysis (from [16], © Wiley InterScience, 2001).

of the EBG structure. This model is simple to understand, but the results are not very accurate because of the simplified approximation of L and C.

The periodic transmission line method is another popularly used technique to analyze EBG structures [16]. Figure 1.6 depicts a transmission line model of EBG structures, where Z_P is the impedance for each periodic element and X_C is the coupling capacitor. The Floquet periodic boundary condition is considered in this approach. After analyzing the cascaded transmission line, the dispersion curve can be readily obtained, which provides more information than the lumped element method. The surface wave modes, leaky wave modes, left- and right-hand regions, and band gaps can be easily identified from the dispersion curve. However, a difficulty in this method is how to accurately obtain the equivalent Z_P and X_C values for the EBG structures. Some empirical formulas have been proposed for simple geometries using multi transmission line (MTL) models, but limited results are found for general geometries.

Owing to the fast development in computational electromagnetics, various numerical methods have been applied in the full wave simulations of EBG structures. Both the frequency domain methods such as the MoM and FEM and the time domain methods like FDTD have been utilized by different research groups to characterize EBG structures. For example, Fig. 1.7 depicts an FDTD model for the mushroom-like EBG analysis [17]. One advantage of the full wave numerical methods is the versatility and accuracy in analyzing different EBG geometries. Another important advantage is the capability to derive various EBG characteristics, such as the surface impedance, reflection phase, dispersion curve, and band gaps. A detailed discussion on the finite difference time domain method will be presented in Chapter 2.

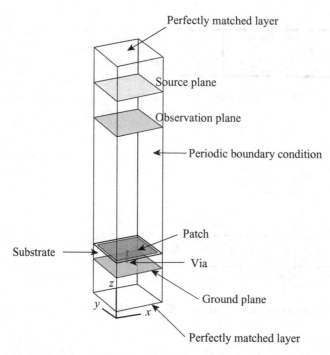

Fig. 1.7 Full wave FDTD model for EBG analysis.

1.4 EBG applications in antenna engineering

The unique electromagnetic properties of EBG structures have led to a wide range of applications in antenna engineering. This section summarizes several typical EBG applications in antenna designs in the hope of stimulating discussions and new avenues of research in this area.

1.4.1 Antenna substrates for surface wave suppressions

Surface waves are by-products in many antenna designs. Directing electromagnetic wave propagation along the ground plane instead of radiation into free space, the surface waves reduce the antenna efficiency and gain. The diffraction of surface waves increases the back lobe radiations, which may deteriorate the signal to noise ratio in wireless communication systems such as GPS receivers. In addition, surface waves raise the mutual coupling levels in array designs, resulting in the blind scanning angles in phased array systems.

The band gap feature of EBG structures has found useful applications in suppressing the surface waves in various antenna designs. For example, an EBG structure is used to surround a microstrip antenna to increase the antenna gain and reduce the back lobe [18–20]. In addition, it is used to replace the quarter-wavelength choke rings in GPS antenna designs [21]. Many array antennas also integrate EBG structures to reduce the

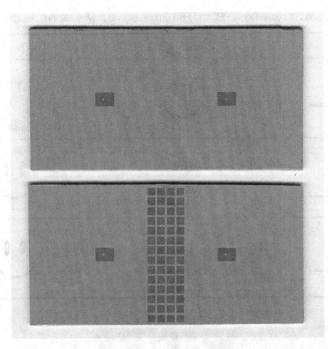

Fig. 1.8 EBG substrate for surface wave suppressions: low mutual coupling microstrip array design (from [22], © IEEE, 2003).

mutual coupling level. For example, Fig. 1.8 shows a comparison of patch antennas with and without EBG structures and an 8 dB reduction in mutual coupling is observed [22].

1.4.2 Antenna substrates for efficient low profile wire antenna designs

Another favorable application of EBG is to design low profile wire antennas with good radiation efficiency, which is desired in modern wireless communication systems. To illustrate the fundamental principle, Table 1.1 compares the EBG with the traditional PEC ground plane in wire antenna designs. When an electric current is vertical to a PEC ground plane, the image current has the same direction and reinforces the radiation from the original current. Thus, this antenna has good radiation efficiency, but suffers from relative large antenna height due to the vertical placement of the current. To realize a low profile configuration, one may position a wire antenna horizontally close to the ground plane. However, the problem is the poor radiation efficiency because the opposite image current cancels the radiation from the original current. In contrast, the EBG surface is capable of providing a constructive image current within a certain frequency band, resulting in good radiation efficiency. In summary, the EBG surface exhibits a great potential for low profile efficient wire antenna applications.

Based on this concept, various wire antennas have been constructed on the EBG ground plane [23–27]. Typical configurations include dipole antenna, monopole antenna, and spiral antenna. EBG surfaces have also been optimized to realize better performance

Table 1.1 Comparisons of conventional PEC and EBG ground planes in wire antenna designs

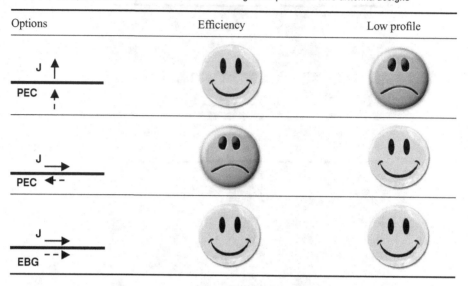

Fig. 1.9 EBG substrate for a low profile curl antenna design (from [24] © Wiley InterScience, 2001).

such as multi-band and wideband designs. For example, Fig. 1.9 shows a curl antenna on an EBG structure that radiates circularly polarized radiation patterns.

1.4.3 Reflection/transmission surfaces for high gain antennas

EBG structures are also applied in designing antennas with a high gain around or above 20 dBi. Traditionally, high gain antennas are realized using either parabolic antennas or large antenna arrays. However, the curved surface of parabolic antennas makes it difficult

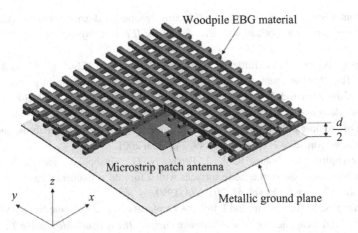

Fig. 1.10 A high gain resonator antenna design using a woodpile EBG structure (from [28], © IEEE, 2005).

for them to be conformal with mobile platforms, while large antenna arrays always suffer from loss in the feeding networks. The planar EBG surfaces provide an alternative solution to this problem. For example, it is used to design a high gain resonator antenna [28], as shown in Fig. 1.10. After optimizing the woodpile EBG design, a 19 dBi antenna gain is obtained in the measurement. Another interesting application is the holographic antenna design [29–31]. Based on the optical holography concept, the antenna operation mechanism is similar to reflectarray antennas. The phase of each individual EBG cell is adjusted in order to produce a narrow high gain beam at a specific direction. Using varactors diodes to reconfigure the surface property, electronic beam steering is also realized [32].

Besides antenna applications, EBG structures have also found numerous applications in microwave circuit designs. A representative example is microwave filter design that successfully rejects the higher harmonics in the circuits. In order to help readers follow the research progress on EBG structures, a literature overview is presented in the Appendix. Readers are encouraged to review this appendix and find more information from related publications.

References

1. B. A. Munk, *Frequency Selective Surfaces: Theory and Design*, John Wiley & Sons, Inc., 2000.
2. J. D. Joannopoulos, R. D. Meade, and J. N. Winn, *Photonic Crystals*, Princeton University Press, 1995.
3. E. Yablonovitch, "Inhibited spontaneous emission in solid-state physics and electronics," *Phys. Rev. Lett.*, **vol. 58**, 2059–63, 1987.

4. Y. Rahmat-Samii and H. Mosallaei, "Electromagnetic band-gap structures: classification, characterization and applications," *Proceedings of IEE-ICAP symposium*, pp. 560–4, April 2001.
5. E. Ozbay, A. Abeyta, G. Tuttle, M. Tringides, R. Biswas, T. Chan, C. M. Soukoulis, and K. M. Ho, "Measurement of a three-dimensional photonic band gap in a crystal structure made of dielectric rods," *Phys. Rev. B, Condens. Matter*, **vol. 50, no. 3**, 1945–8, July 1994.
6. A. S. Barlevy and Y. Rahmat-Samii, "Characterization of electromagnetic band-gaps composed of multiple periodic tripods with interconnecting vias: concept analysis, and design," *IEEE Trans. Antennas Propagat.*, **vol. 49**, 242–353, 2001.
7. D. Sievenpiper, L. Zhang, R. F. J. Broas, N. G. Alexopolus, and E. Yablonovitch, "High-impedance electromagnetic surfaces with a forbidden frequency band," *IEEE Trans. Microwave Theory Tech.*, **vol. 47**, 2059–74, 1999.
8. F.-R. Yang, K.-P. Ma, Y. Qian, and T. Itoh, "A uniplanar compact photonic-bandgap (UC-PBG) structure and its applications for microwave circuit," *IEEE Trans. Microwave Theory Tech.*, **vol. 47**, 1509–14, 1999.
9. V. Radisic, Y. Qian, R. Coccioli, and T. Itoh, "Novel 2-D photonic bandgap structure for microstrip lines," *IEEE Microw. and Guided Wave Lett.*, **vol. 8, no. 2**, 69–71, 1998.
10. C. Caloz and T. Itoh, *Electromagnetic Metamaterials: Transmission Line Theory and Microwave Applications*, Wiley-IEEE Press, 2005.
11. Special issue on electromagnetic crystal structures, designs, synthesis, and applications, *IEEE Trans. Microwave Theory Tech.* **vol. 47, no. 11**, November 1999.
12. Special issue on meta-materials, *IEEE Trans. Antennas Propag.*, **vol. 51, no. 10**, 2003.
13. N. Engheta and R. Ziolkowski, *Metamaterials: Physics and Engineering Explorations*, John Wiley & Sons Inc., 2006.
14. G. V. Eleftheriades and K. G. Balmain, *Negative Refraction Metamaterials: Fundamental Principles and Applications*, Wiley-IEEE Press, 2005.
15. D. F. Sievenpiper, *High Impedance electromagnetic surfaces*, Ph.D. dissertation, Electrical Engineering Department, University of California, Los Angeles, 1999.
16. M. Rahman and M. A. Stuchly, "Transmission line – periodic circuit representation of planar microwave photonic bandgap structures," *Microwave Optical Tech. Lett.*, **vol. 30, no. 1**, 15–19, 2001.
17. Y. Kim, F. Yang, and A. Elsherbeni, "Compact artificial magnetic conductor designs using planar square spiral geometry," *Progress In Electromagnetics Research*, PIER 77, 43–54, 2007.
18. R. Coccioli, F. R. Yang, K. P. Ma, and T. Itoh, "Aperture-coupled patch antenna on UC-PBG substrate," *IEEE Trans. Microwave Theory Tech.*, **vol. 47**, 2123–30, 1999.
19. R. Gonzalo, P. Maagt, and M. Sorolla, "Enhanced patch-antenna performance by suppressing surface waves using photonic-bandgap substrates," *IEEE Trans. Microwave Theory Tech.*, **vol. 47**, 2131–8, 1999.
20. J. S. Colburn and Y. Rahmat-Samii, "Patch antennas on externally perforated high dielectric constant substrates," *IEEE Trans. Antennas Propagat.*, **vol. 47**, 1785–94, 1999.
21. W. E. McKinzie III, R. B. Hurtado, B. K. Klimczak, J. D. Dutton, "Mitigation of multipath through the use of an artificial magnetic conductor for precision GPS surveying antennas," in *Proc. IEEE APS Dig.*, vol. 4, pp. 640–3, 2002.
22. F. Yang and Y. Rahmat-Samii, "Microstrip antennas integrated with electromagnetic band-gap (EBG) structures: a low mutual coupling design for array applications," *IEEE Trans. Antennas Propagat*, **vol. 51, no. 10**, part 2, 2936–46, 2003.

23. Z. Li and Y. Rahmat-Samii, "PBG, PMC and PEC ground planes: A case study for dipole antenna," *IEEE APS Int. Symp. Dig.*, vol. 4, pp. 2258–61, Salt Lake City, UT, July 16–21, 2000.
24. F. Yang and Y. Rahmat-Samii, "A low profile circularly polarized curl antenna over electromagnetic band-gap (EBG) surface," *Microwave Optical Tech. Lett.*, **vol. 31, no. 4**, 264–7, November 2001.
25. F. Yang and Y. Rahmat-Samii, "Reflection phase characterizations of the EBG ground plane for low profile wire antenna applications," *IEEE Trans. Antennas Propagat.*, **vol. 51, no. 10**, 2691–703, 2003.
26. S. Clavijo, R. E. Diaz, and W. E. McKinzie, "Design methodology for Sievenpiper high-impedance surfaces: an artificial magnetic conductor for positive gain electrically small antennas," *IEEE Trans. Antennas Propagat.*, **vol. 51, no. 10**, 2678–90, 2003.
27. H. Nakano, K. Hitosugi, N. Tatsuzawa, D. Togashi, H. Mimaki, and J. Yamauchi, "Effects on the radiation characteristics of using a corrugated reflector with a helical antenna and an electromagnetic band-gap reflector with a spiral antenna," *IEEE Trans. Antennas Propagat.*, **vol. 53, no. 1**, 191–9, 2005.
28. A. R. Weily, L. Horvath, K. P. Esselle, B. C. Sanders, and T. S. Bird, "A planar resonator antenna based on a woodpile EBG material," *IEEE Trans. Antennas Propagat.*, **vol. 53, no. 1**, 216–23, 2005.
29. D. Sievenpiper, J. Colburn, B. Fong, J. Ottusch, J. Visher, "Holographic artificial impedance surfaces for conformal antennas," *IEEE APS Int. Symp. Dig.*, vol. 1B, pp. 256–9, 2005.
30. J. S. Colburn, D. F. Sievenpiper, B. H. Fong, J. J. Ottusch, J. L. Visher, and P. R. Herz, "Advances in artificial impedance surface conformal antennas," *IEEE APS Int. Symp. Dig.*, pp. 3820–3, 2007.
31. F. Caminita, M. Nannetti, and S. Maci, "A new concept for the design of holographic antennas," 2007 URSI Electromagnetic Theory Symposium, July 2007.
32. D. Sievenpiper, J. Schaffner, R. Loo, G. Tangonan, S. Ontiveros, and R. Harold, "A tunable impedance surface performing as a reconfigurable beam steering reflector," *IEEE Trans. Antennas Propagat.*, **vol. 50, no. 3**, 384–90, 2003.

2 FDTD method for periodic structure analysis

2.1 FDTD fundamentals

2.1.1 Introduction

A fundamental quest in electromagnetics and antenna engineering is to solve Maxwell's equations under various specific boundary conditions. In the last several decades, computational electromagnetics has progressed rapidly because of the increased popularity and enhanced capability of computers. Various numerical techniques have been proposed to solve Maxwell's equations [1]. Some of them deal with the integral form of Maxwell's equations while others handle the differential form. In addition, Maxwell's equations can be solved either in the frequency domain or time domain depending on the nature of applications. The success in computational electromagnetics has propelled modern antenna engineering developments.

Among various numerical techniques, the finite difference time domain (FDTD) method has demonstrated desirable and unique features for analysis of electromagnetic structures [2]. It simply discretizes Maxwell's equations in the time and space domains, and electromagnetics behavior is obtained through a time evolving process. A significant advantage of the FDTD method is the versatility to solve a wide range of microwave and antenna problems. It is flexible enough to model various media, such as conductors, dielectrics, lumped elements, active devices, and dispersive materials. Another advantage of the FDTD method is the capability to provide a broad band characterization in one single simulation. Since this method is carried out in the time domain, a wide frequency band response can be obtained through the Fourier transformation of the transient data.

Because of these advantages, the FDTD method has been widely used in many electromagnetic applications. These include the traditional EM problems, such as scattering and radar cross section (RCS) calculations, antenna impedance and pattern characterizations, and microwave circuit designs. The FDTD method also demonstrates a strong capability in many new emerging scenarios, such as wave interaction with human bodies and electromagnetic interference (EMI) in high speed electronics, which are difficult to solve using other methods. In this chapter, we will focus on the analysis of the periodic structure using the FDTD method.

2.1.2 Yee's cell and updating scheme

As a starting point, the FDTD method deals with the *differential form* of the *time domain* Maxwell's equations:

$$\nabla \times \vec{E} = -\frac{\partial \vec{B}}{\partial t}$$
$$\nabla \times \vec{H} = \frac{\partial \vec{D}}{\partial t} + \vec{J} \quad (2.1)$$
$$\nabla \cdot \vec{D} = \rho$$
$$\nabla \cdot \vec{B} = 0,$$

where $\vec{E}$ is the electric field intensity in volts per meter, $\vec{D}$ is the electric flux density in coulombs per square meter, $\vec{H}$ is the magnetic field intensity in amperes per meter, $\vec{B}$ is the magnetic flux density in webers per square meter, $\vec{J}$ represents the electric current density in amperes per square meter, and ρ denotes the electric charge density in coulombs per cubic meter.

Constitutive relations are a necessary supplement to Maxwell's equations, which describe the electric and magnetic characteristics of materials. In a linear, isotropic, and non-dispersive material, they can be simply written as:

$$\vec{D} = \varepsilon \vec{E}, \quad \vec{B} = \mu \vec{H}, \quad \vec{J} = \sigma \vec{E}, \quad (2.2)$$

where ε is the permittivity, μ is the permeability, and σ is the conductivity. In complex media such as chiral materials and artificial absorbing layers, (ε, μ) could be three-dimensional tensors with complex values. For dispersive medium, these constitutive parameters are functions of frequency.

Although there are four equations in (2.1), they are not independent of each other. Actually, the two divergence equations can be derived from the curl equations. Therefore, only the two curl equations needs to be considered in the FDTD method.

To solve Maxwell's equations numerically, Yee introduced a cubic lattice to discretize the computational domain [3]. The cell dimensions are $(\Delta x, \Delta y, \Delta z)$ in Cartesian coordinates. The space domain is then completely filled with these unit cells, and a grid point (i, j, k) is defined as:

$$(i, j, k) = (i \Delta x, j \Delta y, k \Delta z). \quad (2.3)$$

As shown in Fig. 2.1, electric fields are sampled in the middle of the cubic edges while magnetic fields are sampled in the center of the cubic surfaces. Thus, each $\vec{E}$ component is surrounded by four circulating $\vec{H}$ components, and each $\vec{H}$ component is surrounded by four circulating $\vec{E}$ components.

In the time domain, the electric field and magnetic field are evaluated with a half time step difference. Assuming that Δt is the time increment, the electric field is sampled at $t = n\Delta t$ while the magnetic field is sampled at $t = (n+1/2)\Delta t$. In summary, the following

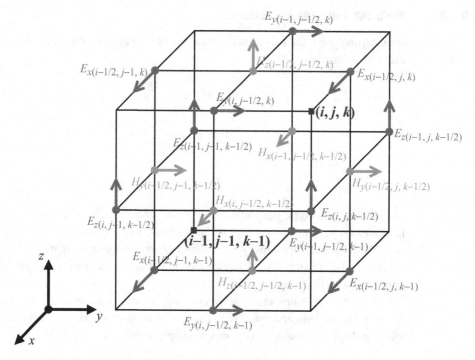

Fig. 2.1 Electric and magnetic field vectors in a Yee's unit cell.

field components are used in later derivations:

$$E^n_{x,i-1/2,j,k}, \quad E^n_{y,i,j-1/2,k}, \quad E^n_{z,i,j,k-1/2} \\ H^{n+1/2}_{x,i,j-1/2,k-1/2}, \quad H^{n+1/2}_{y,i-1/2,j,k-1/2}, \quad H^{n+1/2}_{z,i-1/2,j-1/2,k}. \quad (2.4)$$

With the above definitions, a central finite difference scheme, which has second-order accuracy, is used to calculate the differentiations in the two curl equations in (2.1). As a result, the following six updating equations are obtained for cells in free space:

$$H^{n+1/2}_{x,i,j-1/2,k-1/2} = H^{n-1/2}_{x,i,j-1/2,k-1/2} - \frac{\Delta t}{\mu_0}\left[\frac{E^n_{z,i,j,k-1/2}-E^n_{z,i,j-1,k-1/2}}{\Delta y} - \frac{E^n_{y,i,j-1/2,k}-E^n_{y,i,j-1/2,k-1}}{\Delta z}\right]$$

$$H^{n+1/2}_{y,i-1/2,j,k-1/2} = H^{n-1/2}_{y,i-1/2,j,k-1/2} - \frac{\Delta t}{\mu_0}\left[\frac{E^n_{x,i-1/2,j,k}-E^n_{x,i-1/2,j,k-1}}{\Delta z} - \frac{E^n_{z,i,j,k-1/2}-E^n_{z,i-1,j,k-1/2}}{\Delta x}\right]$$

$$H^{n+1/2}_{z,i-1/2,j-1/2,k} = H^{n-1/2}_{z,i-1/2,j-1/2,k} - \frac{\Delta t}{\mu_0}\left[\frac{E^n_{y,i,j-1/2,k}-E^n_{y,i-1,j-1/2,k}}{\Delta x} - \frac{E^n_{x,i-1/2,j,k}-E^n_{x,i-1/2,j-1,k}}{\Delta y}\right]$$

$$(2.5)$$

$$E^{n+1}_{x,i-1/2,j,k} = E^n_{x,i-1/2,j,k} + \frac{\Delta t}{\varepsilon_0}\left[\frac{H^{n+1/2}_{z,i-1/2,j+1/2,k}-H^{n+1/2}_{z,i-1/2,j-1/2,k}}{\Delta y} - \frac{H^{n+1/2}_{y,i-1/2,j,k+1/2}-H^{n+1/2}_{y,i-1/2,j,k-1/2}}{\Delta z}\right]$$

$$E^{n+1}_{y,i,j-1/2,k} = E^n_{y,i,j-1/2,k} + \frac{\Delta t}{\varepsilon_0}\left[\frac{H^{n+1/2}_{x,i,j-1/2,k+1/2}-H^{n+1/2}_{x,i,j-1/2,k-1/2}}{\Delta z} - \frac{H^{n+1/2}_{z,i+1/2,j-1/2,k}-H^{n+1/2}_{z,i-1/2,j-1/2,k}}{\Delta x}\right]$$

$$E^{n+1}_{z,i,j,k-1/2} = E^n_{z,i,j,k-1/2} + \frac{\Delta t}{\varepsilon_0}\left[\frac{H^{n+1/2}_{y,i+1/2,j,k-1/2}-H^{n+1/2}_{y,i-1/2,j,k-1/2}}{\Delta x} - \frac{H^{n+1/2}_{x,i,j+1/2,k-1/2}-H^{n+1/2}_{x,i,j-1/2,k-1/2}}{\Delta y}\right]$$

$$(2.6)$$

FDTD method is implemented to analyze different types of objects encountered in electromagnetic problems. The updating equations (2.5) and (2.6) can be easily modified to model dielectrics, conductors, thin wires, and lumped elements [4]. In addition, dispersive and non-linear materials can be also accurately characterized with proper models in the FDTD algorithm [5].

In order to calculate the time evolution of electromagnetic fields, the updating equations (2.5) and (2.6) are used in the following manner. At $t = 0$, all electric and magnetic field values are initialized with zero value. A time domain excitation signal is then introduced into the computation domain. For an antenna problem, the excitation can be a lumped voltage source at a specified location (i_0, j_0, k_0). For a scattering problem, a distributed plane wave source is incorporated on a given surface.

Next, the values of magnetic fields are computed at time $t = (n + 0.5)\Delta t$, followed by the computation of electric fields at time $t = (n + 1)\Delta t$. Note that the field values need to be updated on all grid points, which are the major consumptions of computational time and memory storage. The time-stepping iterative procedure is repeated until the desired time response for the electromagnetic problem is obtained. For an antenna or scattering problem, the field values decay to zero as time increases because all energy radiates to infinity. For eigen-value problems such as guided wave analysis, the FDTD computation is terminated when a stable resonant signal is observed.

Finally, a Fourier transform is performed on the time domain data to acquire the frequency characteristics of the electromagnetic problem. According to the Nyquist–Shannon sampling theorem, the maximum frequency is determined by the FDTD time step size. Besides Fourier transformation, some signal processing techniques have also been used to determine the frequency response. For examples, Prony technique and the auto-regressive moving average (ARMA) estimator are used to find the frequency behavior with only a fraction of the time domain data. The system identification technique is used to find the eigen-frequency accurately.

It is worthwhile to emphasize that the aforementioned time-stepping procedure is fully explicit. Thus, no matrix inversion is needed in the FDTD method. As a comparison, matrix inversion is necessary in some other numerical techniques such as the method of moments (MoM). When the number of unknowns increases, a large matrix may become ill-conditioned and its inversion can be troublesome. Therefore, the avoidance of the matrix inversion is a distinguished advantage of the FDTD method.

There are several mathematical issues related to the updating equations (2.5) and (2.6), and the one discussed here is the stability constraint. As an explicit finite difference scheme, it is required that the time step size Δt must be smaller than a specific bound determined by the lattice space increments Δx, Δy, and Δz in order to avoid numerical instability. For three-dimensional problems, the Courant–Friedrich–Lewy (CFL) stability condition gives the following bound for the time step size:

$$\Delta t \leq \frac{1}{c\sqrt{\frac{1}{\Delta x^2} + \frac{1}{\Delta y^2} + \frac{1}{\Delta z^2}}}, \tag{2.7}$$

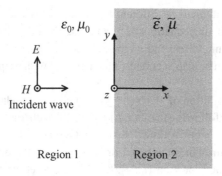

Fig. 2.2 A plane wave normally incident on an interface between two media.

where c is the speed of light. Recently, some enhanced FDTD techniques, such as ADI-FDTD method [6–7], have been developed to relax this constraint. However, they need to modify the updating equations (2.5) and (2.6) with some implicit algorithms.

2.1.3 Absorbing boundary conditions: PML

In many antenna and scattering problems, radiated and scattered fields propagate to infinity. Thus, the computational domain must theoretically extend to infinity for perfect representation of the actual electromagnetics scenario. However, in practice the computational domain must be truncated into a finite size due to restricted computational resources. When the radiated or scattered fields arrive at the boundary, proper boundary conditions need to be implemented to eliminate reflected fields. As a result, the simulation seems to be performed on a computational domain of infinite extent.

Various absorbing boundary conditions (ABC) are developed for this purpose. Modern ABCs have excellent capabilities for virtually reflection-free truncation of the computation domain with a wide dynamic range of 70 dB or more. These absorbing boundary conditions can be categorized into two groups. The first one is a radiation boundary condition that is based on the traveling wave equations [8–9]. Another type of absorbing boundary condition is known as perfectly matched layers (PML) [10–11]. This technique is based on the use of artificial layers appropriately designed to absorb the electromagnetic waves without noticeable reflection. Here, we focus on the PML technique because of its robustness, versatility, and superior performance.

One-dimensional (1-D) perfectly matched layers

As a simple example, the one-dimensional boundary condition is studied first to understand the PML concept. Figure 2.2 shows a plane wave propagating from the free space into a lossy dielectric medium. The constitutive parameters of the dielectric medium are denoted by $(\tilde{\varepsilon}, \tilde{\mu})$ as below:

$$\tilde{\varepsilon} = \varepsilon + \frac{\sigma^e}{j\omega}, \quad \tilde{\mu} = \mu + \frac{\sigma^m}{j\omega}, \tag{2.8}$$

where ε is the permittivity, μ is the permeability, σ^e is the electric conductivity, and σ^m is the magnetic conductivity.

Assume the fields of the incident waves are:

$$\vec{E}^i = \hat{y} E_0 e^{-j\beta_0 x},$$
$$\vec{H}^i = \hat{z} \frac{E_0}{\eta_0} e^{-j\beta_0 x} \tag{2.9}$$
$$\eta_0 = \sqrt{\mu_0/\varepsilon_0}, \quad \beta_0 = \omega\sqrt{\varepsilon_0 \mu_0}.$$

When it impinges the interface, part of wave is reflected back and part transmitted into Region 2. The reflected fields in Region 1 and transmitted field in Region 2 are written as:

$$\vec{E}^r = \hat{y} \Gamma E_0 e^{-j\beta_0 x}, \quad \vec{H}^r = -\hat{z} \Gamma \frac{E_0}{\eta_0} e^{-j\beta_0 x} \tag{2.10}$$

$$\vec{E}^t = \hat{y} \tau E_0 e^{-j\beta_1 x}, \quad \vec{H}^t = \hat{z} \tau \frac{E_0}{\eta_1} e^{-j\beta_1 x} \tag{2.11}$$
$$\eta_1 = \sqrt{\tilde{\mu}/\tilde{\varepsilon}}, \quad \beta_1 = \omega\sqrt{\tilde{\varepsilon}\tilde{\mu}}$$

where Γ is the reflection coefficient and τ is the transmission coefficient. Applying the boundary conditions on the interface, one can obtain the reflection coefficient:

$$\Gamma = \frac{\eta_0 - \eta_1}{\eta_0 + \eta_1} = \frac{\sqrt{\frac{\mu_0}{\varepsilon_0}} - \sqrt{\frac{\mu}{\varepsilon} \cdot \frac{1 + \sigma^m/j\omega\mu}{1 + \sigma^e/j\omega\varepsilon}}}{\sqrt{\frac{\mu_0}{\varepsilon_0}} + \sqrt{\frac{\mu}{\varepsilon} \cdot \frac{1 + \sigma^m/j\omega\mu}{1 + \sigma^e/j\omega\varepsilon}}}. \tag{2.12}$$

If we set $\varepsilon = \varepsilon_0$ and $\mu = \mu_0$ and further enforce the condition

$$\frac{\sigma^m}{\mu} = \frac{\sigma^e}{\varepsilon}, \tag{2.13}$$

then $\eta_1 = \eta_0$ and the reflection coefficient $\Gamma = 0$. Thus, a reflectionless interface between Region 1 and Region 2 is obtained. Region 2 is said to be *perfectly matched* to Region 1 under the normal incidence condition.

With the constitutive parameters set above, the transmitted fields in Region 2 can be written as:

$$\eta_1 = \eta_0, \quad \beta_1 = \beta_0 - j\sigma^e \eta_0$$
$$\vec{E}^t = \hat{y} E_0 e^{-j\beta_0 x} e^{-\sigma^e \eta_0 x} \tag{2.14}$$
$$\vec{H}^t = \hat{z} \frac{E_0}{\eta_0} e^{-j\beta_0 x} e^{-\sigma^e \eta_0 x}.$$

It is noticed that the transmitted wave in Region 2 attenuates exponentially along the propagation direction x. After a certain distance, the field strength approaches zero. Therefore, the computation domain in the FDTD simulation can be truncated naturally. In summary, *reflectionless on the interface* and *attenuation in the lossy medium* are the two key points in PML design.

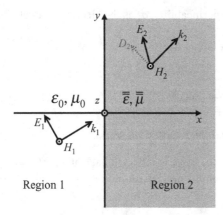

Fig. 2.3 A TM plane wave incident from free space to a uniaxial medium.

When the above PML concept is used to develop FDTD code, some numerical errors are observed. One error results from the finite thickness truncation of the PML, which can be minimized by increasing conductivity value. Another one is due to the discontinuity of the conductivities in the two media. When the E field on the interface between free space and PML is updated, an average conductivity is used. This numerical approximation incurs a noticeable error in the FDTD simulation, which can be minimized by setting a small conductivity near the interface.

To reduce both numerical errors, a non-uniform PML technique is proposed [12]. The idea is to start the PML with zero conductivity at the interface and gradually increase its value as the PML depth increases. For example, a polynomial distribution is used to set up the conductivity:

$$\sigma^e(x) = (x/l)^m \sigma^e_{\max}. \tag{2.15}$$

The value of the conductivity increases from 0 at the interface ($x = 0$) to $\sigma^e_{\max}$ at the truncation boundary ($x = l$). Hence, both finite thickness truncation error and discontinuous conductivity error can be suppressed to a desired level.

Three-dimensional (3-D) perfectly matched layers

The PML concept is extended to solve three-dimensional problems. For general 3-D applications, a difficulty in designing PML is how to absorb electromagnetic waves at *all incident angles*, *all frequencies*, and *all polarizations*. This problem is solved by introducing tensor-form permittivity and permeability.

To demonstrate the basic principle, let's consider a TM wave obliquely incident from free space to a uniaxial medium, as shown in Fig. 2.3. The permittivity and permeability of the uniaxial medium are defined in tensor forms as below:

$$\bar{\bar{\varepsilon}} = \varepsilon_0 \begin{bmatrix} a & 0 & 0 \\ 0 & b & 0 \\ 0 & 0 & b \end{bmatrix}, \quad \bar{\bar{\mu}} = \mu_0 \begin{bmatrix} c & 0 & 0 \\ 0 & d & 0 \\ 0 & 0 & d \end{bmatrix}. \tag{2.16}$$

In the isotropic region 1, the fields are expressed as a superposition of the incident and reflected fields as:

$$\vec{H}_1 = \hat{z} H_0 (1 + \Gamma e^{2jk_{1x}x}) e^{-jk_{1x}x - jk_{1y}y}$$

$$\vec{E}_1 = \left[\hat{x} \frac{-k_{1y}}{\omega \varepsilon_0} H_0 (1 + \Gamma e^{2jk_{1x}x}) + \hat{y} \frac{k_{1x}}{\omega \varepsilon_0} H_0 (1 - \Gamma e^{2jk_{1x}x}) \right] e^{-jk_{1x}x - jk_{1y}y} \quad (2.17)$$

$$k_{1x}^2 + k_{1y}^2 = \omega^2 \varepsilon_0 \mu_0.$$

The transmitted wave in Region 2 is also a TM wave and the fields are expressed as [12]:

$$\vec{H}_2 = \hat{z} \tau H_0 e^{-jk_{2x}x - jk_{2y}y}$$

$$\vec{E}_2 = \left(-\hat{x} \frac{k_{2y}}{\omega \varepsilon a} + \hat{y} \frac{k_{2x}}{\omega \varepsilon b} \right) \tau H_0 e^{-jk_{2x}x - jk_{2y}y} \quad (2.18)$$

$$\frac{k_{2x}^2}{bd} + \frac{k_{2y}^2}{ad} = \omega^2 \varepsilon_0 \mu_0.$$

Here, Γ and τ are reflection and transmission coefficients. Applying the boundary conditions on the interface, we obtain:

$$1 + \Gamma = \tau$$

$$\frac{k_{1x}}{\omega \varepsilon_0} (1 - \Gamma) = \frac{k_{2x}}{\omega \varepsilon b} \tau \quad (2.19)$$

$$k_{1y} = k_{2y}.$$

Solving the above equations, the reflection and transmission coefficients are obtained:

$$\Gamma = \frac{k_{1x} - k_{2x}/b}{k_{1x} + k_{2x}/b}, \quad \tau = 1 + \Gamma = \frac{2k_{1x}}{k_{1x} + k_{2x}/b}. \quad (2.20)$$

In order to realize the reflectionless condition ($\Gamma = 0$), the following equation must be satisfied:

$$k_{1x} = k_{2x}/b. \quad (2.21)$$

Considering the dispersion relations in (2.17) and (2.18), the phase matching condition in (2.19), and the reflectionless condition in (2.21), we obtain:

$$\omega^2 \varepsilon_0 \mu_0 \left(1 - \frac{d}{b} \right) = k_{1y}^2 \left(1 - \frac{1}{ab} \right). \quad (2.22)$$

Since the PML is desired to work at all frequencies and all incident angles, the above equation should be satisfied for arbitrary ω and k_{1y}. Therefore, the constitutive parameters must be set as:

$$d = b, \quad ab = 1. \quad (2.23)$$

The above procedure is repeated for the TE incidence, and the following constitutive relations are derived to realize zero reflection:

$$d = b, \quad cd = 1. \quad (2.24)$$

Combining the results in (2.23) and (2.24), we can design the uniaxial medium with following permittivity and permeability tensors so that it will be reflectionless for both

TE^z and TM^z polarizations:

$$\bar{\bar{\varepsilon}} = \varepsilon_0 \bar{\bar{s}}, \quad \bar{\bar{\mu}} = \mu_0 \bar{\bar{s}}, \quad \bar{\bar{s}} = \begin{bmatrix} s_x^{-1} & 0 & 0 \\ 0 & s_x & 0 \\ 0 & 0 & s_x \end{bmatrix}. \tag{2.25}$$

Note this setup is for an interface that is parallel to the yz plane with a normal in the x direction.

Similar to the 1-D case, propagation loss needs to be added in order to attenuate the wave in the absorbing medium. Let's choose $s_x = 1 + \frac{\sigma_x}{j\omega\varepsilon_0}$. Then from (2.21), we have,

$$k_{2x} = k_{1x} s_x = k_{1x} - j\eta_0 \cos\theta \sigma_x, \tag{2.26}$$

where θ is the incident angle. Substituting (2.26) into (2.18), the magnetic field in the lossy uniaxial medium is:

$$\vec{H}_2 = \hat{z}\, H_0 e^{-jk_{1x}x - jk_{1x}y} e^{-\sigma_x x \eta_0 \cos\theta}. \tag{2.27}$$

As expected, the field decays exponentially as it propagates along the x direction.

The above derivation for the yz plane interface is repeated for interfaces located on xy and xz planes. As a result, a generalized constitutive tensor $\bar{\bar{s}}$ is obtained as follows:

$$\bar{\bar{s}} = \begin{bmatrix} s_x^{-1} & 0 & 0 \\ 0 & s_x & 0 \\ 0 & 0 & s_x \end{bmatrix} \cdot \begin{bmatrix} s_y & 0 & 0 \\ 0 & s_y^{-1} & 0 \\ 0 & 0 & s_y \end{bmatrix} \cdot \begin{bmatrix} s_z & 0 & 0 \\ 0 & s_z & 0 \\ 0 & 0 & s_z^{-1} \end{bmatrix} \tag{2.28}$$

$$s_x = 1 + \frac{\sigma_x}{j\omega\varepsilon_0}, \quad s_y = 1 + \frac{\sigma_y}{j\omega\varepsilon_0}, \quad s_z = 1 + \frac{\sigma_z}{j\omega\varepsilon_0}. \tag{2.29}$$

For PML along $\pm x$ axes, the conductivities are set as follows: $\sigma_y = \sigma_z = 0$, σ_x is progressively increasing along $\pm x$ axes, as shown in (2.15). Similarly, ($\sigma_x = 0$, σ_y, $\sigma_z = 0$) and ($\sigma_x = 0$, $\sigma_y = 0$, σ_z) are assigned progressively increasing for PML walls along $\pm y$ axes and $\pm z$ axes, respectively.

2.1.4 FDTD excitation

A proper excitation is needed for FDTD simulations. Various excitation models have been proposed in different applications. For antenna designs, a simple voltage source is usually used:

$$V_s^n = V_s(n\Delta t) = -E_{z,i,j,k-1/2}^n \Delta z, \tag{2.30}$$

where the source $(i, j, k - 1/2)$ is located at the antenna feed point. This gap source has been used in antenna characterizations such as dipole antenna and monopole antenna. In addition, the coaxial probe model [13] and the microstrip line model [14] have also been developed for antenna simulations. All these models are local sources, where the excitation is added at the feed location.

Another type of source is the distributed source, where the excitation signal is incorporated on a specified surface. A representative example is the plane wave source for scattering analysis [15]. In this source setup, a virtual incident surface is specified in the computational domain, which separates the computational domain into the total field region and scattered field region. On the virtual surface, the tangential components of incident fields are incorporated. For example, if a virtual incident surface is located at $z = z_0 = k_0 \Delta z$, the following equations are used after the updating equations (2.5) and (2.6) to excite a plane wave:

$$H^{n+1/2}_{x,i,j-1/2,k_0+1/2} = H^{n+1/2}_{x,i,j-1/2,k_0+1/2} + \frac{\Delta t}{\mu_0 \Delta z} E^{\text{incident},n}_{y,i,j-1/2,k_0}$$

$$H^{n+1/2}_{y,i-1/2,j,k_0+1/2} = H^{n+1/2}_{y,i-1/2,j,k_0+1/2} - \frac{\Delta t}{\mu_0 \Delta z} E^{\text{incident},n}_{x,i-1/2,j,k_0} \quad (2.31)$$

$$E^{n+1}_{x,i-1/2,j,k_0} = E^{n+1}_{x,i-1/2,j,k_0} - \frac{\Delta t}{\varepsilon_0 \Delta z} H^{\text{incident},n+1/2}_{y,i-1/2,j,k_0+1/2}$$

$$E^{n+1}_{y,i,j-1/2,k_0} = E^{n+1}_{y,i,j-1/2,k_0} + \frac{\Delta t}{\varepsilon_0 \Delta z} H^{\text{incident},n+1/2}_{x,i,j-1/2,k_0+1/2}.$$

In both the local source and distributed source, the time domain signal should be carefully designed to cover the frequency band of interest. A popular one is a modulated Gaussian waveform:

$$f(t) = \exp\left[-\frac{(t-t_0)^2}{2\sigma_t^2}\right] \cos(2\pi f_0 t). \quad (2.32)$$

The parameters of this modulated Gaussian waveform are the time delay t_0, center frequency f_0, and pulse width σ_t. The time delay t_0 is set to a value between $3\sigma_t$ to $5\sigma_t$ for a smooth start in time domain excitation. The pulse width σ_t determines the frequency bandwidth of the signal. A small σ_t gives a narrow pulse in the time domain but a wide bandwidth in the frequency domain. For convenience, a frequency bandwidth (BW) is defined as below:

$$\text{BW} \times 2\pi \equiv 2 \cdot \frac{3}{\sigma_t}. \quad (2.33)$$

From the definition above, the signal magnitude at $f_0 \pm (\text{BW}/2)$ is 40 dB lower than the signal level at the center frequency f_0.

2.1.5 Extraction of characteristic parameters

At the end of FDTD simulations, characteristic parameters are extracted to describe the electromagnetic behavior of the analyzed object. For antenna designs, input impedance and radiation pattern are required. For scattering analysis, transmission and reflection coefficients are of interest. In a waveguide problem, the eigen-frequency needs to be identified.

The extraction procedure varies with applications. Here, we explain the derivation of the reflection coefficient, which will be used later for scattering analysis of periodic structures. An observational plane is set up at $z = z_1 = k_1 \Delta z$. The incident and scattered fields on this surface are summed together to get the equivalent far field information [2].

If the co-polarized field is along the y direction, following equations are used:

$$E_y^{\text{scatter}} = \frac{1}{N_x N_y} \sum_{i=1}^{N_x} \sum_{j=1}^{N_y} E_y^{\text{scatter}}(i, j - 1/2, k_1) e^{jk_x(i\Delta x)}$$

$$E_y^{\text{incident}} = \frac{1}{N_x N_y} \sum_{i=1}^{N_x} \sum_{j=1}^{N_y} E_y^{\text{incident}}(i, j - 1/2, k_1) e^{jk_x(i\Delta x)}.$$

(2.34)

Note that the proper phase delay is inserted to obtain the correct far field value. The reflection coefficient is then calculated as following:

$$\Gamma = \frac{E_y^{\text{scatter}}}{E_y^{\text{incident}}}.$$

(2.35)

2.2 Periodic boundary conditions

For an infinite structure, it is impossible to directly simulate it using limited computation resources. Thus, the computation domain needs to be truncated using proper boundary conditions. The perfectly matched layers (PML) discussed in the previous section are used to absorb radiating energy from antennas or scatterers that have *finite sizes*. In many applications, especially in artificial electromagnetic materials, an EM structure itself extends to *infinity* in a periodic manner. Typical examples include corrugated waveguides, frequency selective surfaces, electromagnetic band gap structures, and double negative materials. In this situation, the PML that absorbs radiating waves are no longer applicable and a new boundary condition needs to be developed to truncate the computational domain. For this purpose, a periodic boundary condition (PBC) that models the effect of periodic replication is introduced to truncate the computational domain so that only a single unit cell needs to be simulated [16].

2.2.1 Fundamental challenges in PBC

All periodic boundary conditions are developed based on the Floquet theory [17]. For a periodic structure with periodicity p along the x direction, electromagnetic fields at two boundaries $x = 0$ and $x = p$ satisfy the following equations in the frequency domain:

$$\vec{E}(x = 0, y, z) = \vec{E}(x = p, y, z) e^{jk_x p}$$
$$\vec{H}(x = 0, y, z) = \vec{H}(x = p, y, z) e^{jk_x p}.$$

(2.36)

The exponential term represents a propagation phase delay, which is determined by the propagation constant k_x and the periodicity p.

For a waveguide problem, the propagation constant k_x can be obtained from the dispersion relation. However, the dispersion relation for a complex structure is usually unknown before the FDTD simulation.

For a scattering problem, the propagation constant k_x is known, which is determined by the frequency and incident angle:

$$k_x = k_0 \cdot \sin\theta = 2\pi f \sqrt{\varepsilon_0 \mu_0} \cdot \sin(\theta)$$

(2.37)

where k_0 is the free space wavenumber. However, when (2.36) is converted to the time domain using the Fourier transformation, one obtains:

$$\vec{E}(x=0,y,z,t) = \vec{E}(x=p,y,z,t+p\sin\theta/c)$$
$$\vec{H}(x=0,y,z,t) = \vec{H}(x=p,y,z,t+p\sin\theta/c) \quad (2.38)$$

where $c = 1/\sqrt{\varepsilon_0\mu_0}$ is the free space wave speed. It is noticed that when updating electric and magnetic fields in the *current time*(t), the field data in the *future time* $(t+p\sin\theta/c)$ are needed. It opposes the causal relation in the time domain simulation.

2.2.2 Overview of various PBCs

In general, periodic boundary condition (PBC) is used for two types of problems: waveguide characterization and scattering analysis. Different PBCs have been developed to solve these problems. For waveguide problems, a popular time domain boundary condition is derived from (2.36) by setting a fixed wavenumber k_x [18–19]:

$$\vec{E}(x=0,y,z,t) = \vec{E}(x=p,y,z,t) \cdot e^{jk_xp}$$
$$\vec{H}(x=0,y,z,t) = \vec{H}(x=p,y,z,t) \cdot e^{jk_xp} \quad (2.39)$$

The eigen-frequencies at this specified wavenumber are identified after FDTD simulation. The FDTD simulation repeats for a series of k_x values to obtain the dispersion curve.

For scattering problems, the wavenumber is a function of the incident angle θ, as shown in (2.37), and different techniques have been proposed to solve the challenge in (2.38). The normal incidence case is the one that was first solved [20]. When a plane wave is normally incident on a periodic structure, we have $\theta = 0$ and $k_x = 0$. Therefore, (2.38) is simplified to:

$$\vec{E}(x=0,y,z,t) = \vec{E}(x=p,y,z,t)$$
$$\vec{H}(x=0,y,z,t) = \vec{H}(x=p,y,z,t). \quad (2.40)$$

It is clear that no data in the future time are needed in this periodic boundary condition. Hence, the difficulty in (2.38) disappears and PBC can be implemented easily.

When a plane wave is obliquely incident on a periodic structure, a sine-cosine technique was proposed [21]. The computation domain is excited by dual plane wave incidences simultaneously: one with a $\cos\omega t$ time dependence and the other with $\sin\omega t$. The corresponding fields are denoted by (E_1, H_1) and (E_2, H_2). The conventional Yee's algorithm is used to update these two sets of fields separately. To apply the PBC, the fields on the boundary are first combined as follows:

$$\vec{E}_c = \vec{E}_1 + j\vec{E}_2, \quad \vec{H}_c = \vec{H}_1 + j\vec{H}_2 \quad (2.41)$$

It is worthwhile to emphasize that the combined fields have complex values. Since the combined fields have the time convention $e^{j\omega t} = \cos\omega t + j\sin\omega t$, the phase relation in (2.36) can be used directly for the combined fields. The k_x value is calculated using (2.37) with the known frequency ω and incident angle θ. After applying PBC, two set of fields are extracted from the combined fields:

$$(\vec{E}_1, \vec{H}_1) = \mathrm{Re}(\vec{E}_c, \vec{H}_c), \quad (\vec{E}_2, \vec{H}_2) = \mathrm{Im}(\vec{E}_c, \vec{H}_c). \quad (2.42)$$

This procedure is repeated in each time step until a steady state is reached. This method successfully circumvents the difficulty in (2.38) but only a single frequency ω is calculated per simulation.

To improve the computation efficiency and obtain a wideband response per simulation, a split-field method [22] is developed based on field transformation technique [23]. To account for the phase shift along the x direction, a set of auxiliary fields $\vec{P}$ and $\vec{Q}$ are introduced as following:

$$\vec{P} = \vec{E} e^{jk_x x}, \quad \vec{Q} = \vec{H} e^{jk_x x}. \tag{2.43}$$

The periodic boundary conditions for $\vec{P}$ and $\vec{Q}$ can be derived from (2.36):

$$\vec{P}(x=0, y, z) = \vec{P}(x=p, y, z)$$
$$\vec{Q}(x=0, y, z) = \vec{Q}(x=p, y, z). \tag{2.44}$$

Obviously, the boundary conditions are simple and easy to implement. The equations for auxiliary fields are derived from Maxwell's equations:

$$\nabla \times \vec{E} = -j\omega\mu\vec{H} \Rightarrow \nabla \times \vec{P} - jk_x\hat{x} \times \vec{P} = -j\omega\mu\vec{Q}$$
$$\nabla \times \vec{H} = j\omega\varepsilon\vec{H} \Rightarrow \nabla \times \vec{Q} - jk_x\hat{x} \times \vec{Q} = j\omega\varepsilon\vec{P}. \tag{2.45}$$

Thus, the conventional leap-frog algorithm is not applicable now. Instead, a split-field updating algorithm is proposed to solve the above equations and details are presented in [22]. A major limitation of this method is the stability constraint. A simple and conservative limit can be made from the maximum phase velocity in the grid:

$$\frac{c\Delta t}{\Delta x} \leq \frac{1 - \sin\theta}{\sqrt{D}}, \tag{2.46}$$

where D is the dimensionality of the problem. For normal incidence, this stability condition is the same as (2.7). When the incident angle increases, the required time step size becomes smaller. If the incident angle is close to the grazing angles such as 85°, the time step is so tiny that the FDTD simulation time will be prohibitively long.

Besides the periodic boundary conditions mentioned above, there are a number of other approaches proposed for periodic structure analysis, such as multiple unit cell method [24], unit cell shifting method [25], angle-updated method [26], etc. They have been used to analyze various electromagnetic objects.

2.2.3 Constant k_x method for scattering analysis

After reviewing the literature, it is noticed that the PBCs for scattering analysis, no matter the sine-cosine method or split-field method, are much more complicated than the PBC for waveguide analysis. In addition, they have some limitations on the computational efficiency or stability constraint. A question arises: can the waveguide PBC in (2.39) also be used for scattering analysis? The answer is "YES" and here we explain how it is utilized in scattering analysis [27–28].

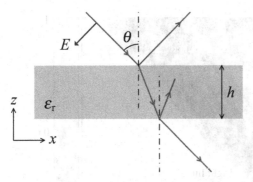

Fig. 2.4 A plane wave impinges upon a dielectric slab (from [35], © American Geophysical Union, 2007). The incident angle of the TM^z wave is denoted by θ. The dielectric slab has a thickness (h) of 9.375 mm and dielectric constant (ε_r) of 2.56.

Reflection coefficient in the k_x-frequency plane

To implement the PBC (2.39) in scattering analysis, the first thing is to have a comprehensive understanding of the k_x-frequency plane. In light of this, a dielectric slab is used as an example to illustrate the reflection coefficient in the k_x-frequency plane. Figure 2.4 shows a TM^z plane wave illuminating on a dielectric slab of thickness h and dielectric constant ε_r.

The reflection coefficient of the dielectric slab is calculated analytically and plotted in the k_x-frequency plane, as shown in Fig. 2.5. The horizontal axis denotes the wavenumber along the x direction (k_x), and the vertical axis represents the frequency. The magnitude of the reflection coefficient (Γ) is represented by gray scales as indicated by the scale bar on the right. The k_x-frequency plane is separated into the guided wave region ($k_x^2 > k_0^2$) and plane wave region ($k_x^2 < k_0^2$) by the light line ($k_x = k_0$). For our discussion on the scattering analysis, only the reflection coefficient in the plane wave region needs to be considered. The normal incidence and an oblique incidence are also indicated by the dashed lines in this k_x-frequency plane.

It is observed that the total transmission, represented by dark scales, occurs in two zones. The first one starts at 10 GHz when k_x equals to zero (normal incidence). The frequency increases to 12 GHz as the horizontal wavenumber increases. The second zone is around a tilted line in the k_x-frequency plane, which corresponds to an incident angle of 58°, the Brewster angle for this case. The reflection coefficient plotted in the k_x-frequency plane provides a complete and clear picture on the scattering property for all incident angles and all frequencies.

Rationale of the constant k_x method

It is instructive to review the previous periodic boundary conditions on the k_x-frequency plane. For a normal incident case, we have $\theta = 0$ and $k_x = 0$. The reflection coefficient is calculated on the leftmost dashed line in Fig. 2.5. The sine-cosine method, which is developed to calculate the reflection coefficient at oblique incidence but only at a single frequency, is indicated by a plus sign. The wideband split-field method calculates the reflection coefficient at a given incident angle, as represented by the tilted dashed line.

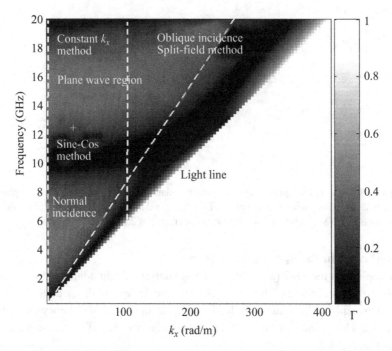

Fig. 2.5 Reflection coefficient of the dielectric slab in the k_x-frequency plane (from [35], © American Geophysical Union, 2007). Various methods to model periodic boundary conditions in FDTD simulations are also indicated in the k_x-frequency plane by the dashed lines and the plus sign.

Alternatively, one can also calculate the reflection coefficient on a vertical line in the k_x-frequency plane like the middle dashed one in Fig. 2.5. On this line, the horizontal wavenumber k_x is a constant. The relationship of the horizontal wavenumber, frequency, and incident angle is described in (2.37). When a constant k_x is chosen, change of frequency leads to correlated change of the incident angle. Therefore, FDTD simulation is performed based on a constant k_x but a variant incident angle θ, which distinguishes the proposed method from the early PBC approaches.

The reason for choosing a constant k_x is recognized from the transformation of (2.36) into the time domain, where a very simple periodic boundary condition can be obtained:

$$\vec{E}(x=0, y, z, t) = \vec{E}(x=p, y, z, t)e^{jk_x p}$$
$$\vec{H}(x=0, y, z, t) = \vec{H}(x=p, y, z, t)e^{jk_x p}.$$
(2.47)

No time delay or advancement is required in this equation. Therefore, it can be directly used to update the field values on the periodic boundary. This PBC is mathematically the same as (2.39), but it has its own physical meaning in scattering problems. Also note that $e^{jk_x p}$ in (2.47) is a complex constant number, resulting in complex E and H field values in the FDTD simulation.

The physical meaning of the constant k_x method can be also understood from the principle of superposition. For a given frequency and a given k_x, time domain equation

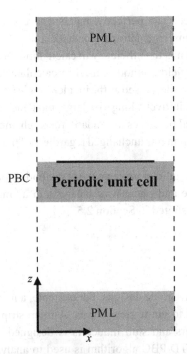

Fig. 2.6 FDTD computational domain of a periodic structure, which is truncated by periodic boundary conditions (PBC) in the horizontal directions and perfectly matched layers (PML) along the z direction.

(2.47) is true for periodic structures. In addition, it is a linear equation for the electric and magnetic fields. Thus, when a wideband pulse is launched, (2.47) still holds by the superposition of all frequency components.

Implementation of the constant k_x method and its advantages

The implementation of this FDTD/PBC algorithm is simple and straightforward. Figure 2.6 shows an FDTD model, where a single unit cell is surrounded by periodic boundary conditions (PBC) in the horizontal directions and perfectly matched layers (PML) along the z direction. The electric and magnetic fields are computed as following:

- In the interior computational domain, the conventional Yee's scheme is used to update the electromagnetic fields, as illustrated in (2.5) and (2.6).
- On the periodic boundaries of the computational domain, (2.47) is used to update the electromagnetic fields.
- In the PML region, (2.28) and (2.29) are used to set up the material property to eliminate the reflection. Note that the periodic boundary also extends into the PML region.

It is worthwhile to highlight several important advantages of the constant k_x method. First, this algorithm is very easy to implement. In contrast to the field transformation methods that use auxiliary fields P and Q, the new approach computes E and H fields directly. Thus, no complicated discretization formulas need to be derived and the traditional Yee's updating scheme is used. The perfectly matched layers also remain unchanged, which helps to simplify the implementation procedure.

Another advantage of the new algorithm is the efficiency in calculating the scattering at large incident angles. When the split-field method is used to calculate the oblique incidence, the time step size (Δt) needs to be decreased as the incident angle (θ) increases. Thus the simulation time becomes prohibitively long for large incidence angles. In contrast, the constant k_x method presented here uses a standard Yee's scheme to update the E and H fields. The stability condition remains unchanged regardless of the horizontal wavenumber (k_x) or incident angle.

Finally, this PBC is consistent with the guided wave analysis method. It provides an opportunity to combine the guided wave and plane wave analysis in a single FDTD simulation. This point will be further illustrated in Section 2.5.

2.3 Guided wave analysis

Periodic structures are used in various waveguide designs. For example, a metal ground plane with periodic corrugations is used to guide slow waves. A microstrip line periodically loaded with inter-digit capacitors and stub inductors is designed to support backward waves. In this section, the FDTD/PBC algorithm is used to analyze several typical waveguides and their dispersion relations are illustrated.

2.3.1 Problem statement

A fundamental question in waveguide analysis is how to accurately characterize the dispersion relation of the structure [29]. Assume an electromagnetic wave propagates along the x direction in a periodic waveguide. *At a given frequency ω, the propagation constants k_x (or β) of guided waves need to be identified.* The results are usually plotted in the k_x-frequency plane, and are known as the dispersion curves. For example, Fig. 2.7 shows the dispersion curves of a dielectric slab waveguide. There are four curves in this figure and each represents a specific surface wave mode. At the frequency 25 GHz, the corresponding propagation constants are 539 radian/m, 575 radian/m, 736 radian/m, and 771 radian/m, respectively. It is observed that cutoff frequencies exist for higher-order modes, such as the TM_1 and TE_1 modes in Fig. 2.7.

An alternative way to state the question is *at a given propagation constant k_x find the corresponding eigen-frequencies of surface waves.* For example, if we set k_x to 600 radian/m in Fig. 2.7, the corresponding eigen-frequencies are 20.03 GHz, 21.22 GHz, 25.74 GHz, and 27.19 GHz, respectively. Since the FDTD method is a time domain method that covers a wideband frequency in a single simulation, it is preferable to use the latter statement as a goal to analyze waveguide problems. When the k_x is specified, the periodic boundary condition (2.39) is directly used in the simulation process. After the time domain simulation arrives at a steady state, the Fourier transform is performed and the eigen-frequencies can be identified from the spikes in the frequency spectrum.

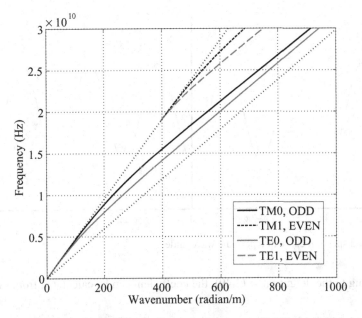

Fig. 2.7 Dispersion curves of a dielectric slab waveguide. The slab thickness is 6.35 mm and the dielectric constant is 2.56.

2.3.2 Brillouin zone for periodic waveguides

A distinguished feature of periodic waveguides is that their dispersion curves are also periodic with respect to k_x [30]. According to Floquet theorem, fields in a periodic structure satisfy the following equations:

$$\vec{E}(x+mp, y, z) = \vec{E}(x, y, z)e^{-jk_x mp}, \tag{2.48}$$

where m is an arbitrary integer, p is the periodicity in the x direction. Assume the wave propagates along the x direction with a wavenumber k_x. If we define an auxiliary field as below:

$$\vec{E}(x, y, z) = \vec{F}(x, y, z)e^{-jk_x x}. \tag{2.49}$$

It can be easily proven that the auxiliary field is a periodic function with respect to x:

$$\vec{F}(x+mp, y, z) = \vec{F}(x, y, z). \tag{2.50}$$

As a periodic function, we can expand it into a Fourier series using the periodicity p:

$$\vec{F}(x, y, z) = \sum_{n=-\infty}^{\infty} \vec{E}_n(y, z)e^{-jn\frac{2\pi}{p}x}. \tag{2.51}$$

The coefficient of each component can be determined from the orthogonality relation:

$$\vec{E}_n(y, z) = \frac{1}{p}\int_0^p \vec{F}(x, y, z)e^{jn\frac{2\pi}{p}x} dx. \tag{2.52}$$

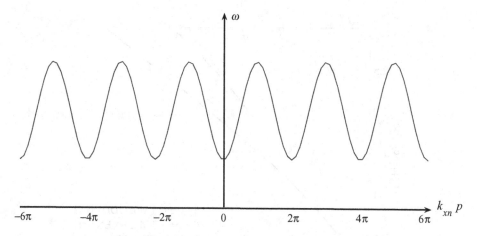

Fig. 2.8 A dispersion curve of a periodic waveguide.

According to the definition in (2.49), the coefficient can be calculated from the $\vec{E}$ field directly:

$$\vec{E}_n(y,z) = \frac{1}{p}\int_0^p \vec{E}(x,y,z)e^{jk_x x}e^{jn\frac{2\pi}{p}x}dx = \frac{1}{p}\int_0^p \vec{E}(x,y,z)e^{jk_{xn}x}dx, \qquad (2.53)$$

where,

$$k_{xn} = k_x + n\frac{2\pi}{p}, \quad n = -\infty, \ldots, -1, 0, 1, \ldots, \infty. \qquad (2.54)$$

Finally, from (2.49) and (2.51), we can express the $\vec{E}$ field also into a series form:

$$\vec{E}(x,y,z) = \sum_{n=-\infty}^{\infty} \vec{E}_n(y,z)e^{-jk_{xn}x}. \qquad (2.55)$$

Thus, the $\vec{E}$ field is a summation of traveling waves. Each traveling wave has a constant magnitude $\vec{E}_n(y,z)$ along the propagation direction x and a corresponding propagation constant k_{xn}. These space harmonic waves have different phase velocities:

$$v_{pn} = \frac{\omega}{k_{xn}} = \frac{\omega}{k_{x0} + n\frac{2\pi}{p}}, \qquad (2.56)$$

but the same group velocity:

$$v_{gn} = \frac{d\omega}{dk_{xn}} = \frac{d\omega}{d\left(k_{x0} + n\frac{2\pi}{p}\right)} = \frac{d\omega}{dk_{x0}} = v_{g0}. \qquad (2.57)$$

Note that for negative *integer* n, traveling waves may have negative phase velocities, e.g., that phase velocity is opposite to the group velocity.

All the space harmonics $\vec{E}_n(y,z)$ share the same time convention $e^{j\omega t}$. In other words, at a given frequency ω, a series of k_x can be found that satisfies (2.54). As a result, a periodic dispersion curve is obtained. For example, Fig. 2.8 sketches a dispersion curve of a periodic structure. Therefore, the periodic waveguide analysis only needs to focus on the dispersion relation within a single period, namely, $0 \leq k_{xn} \leq 2\pi/p$, which is known as the Brillouin zone.

2.3.3 Examples

A dielectric slab waveguide

To demonstrate the validity of FDTD/PBC method in guided wave analysis, two examples are studied here. The first one is a dielectric slab waveguide whose analytical solution is readily derived. The slab has a thickness of 6.35 mm and a dielectric constant of 2.56. Figure 2.7 shows the theoretic dispersion curves of two TE modes and two TM modes.

Although this structure is uniform in the propagation direction, it can be regarded as a periodic structure with an infinitely small periodicity. Thus, the developed FDTD/PBC method can still be used to analyze its behavior. Along the propagation direction, only one cell is selected to increase the computation efficiency.

To apply the periodic boundary condition (2.39), the propagation constant k_x needs to be specified before FDTD simulation. For instance, let us choose the k_x value to be 300 radian/m. A gap source is incorporated to excite the structure and its polarization is consistent with the polarization of guided wave modes. The gap source is added to the E_x component for the TM_x mode and to the H_x component for the TE_x mode. When the FDTD simulation reaches a steady state, the frequency spectrum is obtained by taking the Fourier transform of the time domain data. In this example, an observation point is situated on the interface between the slab and free space. The frequency spectrum of TM mode and TE mode are plotted in Fig. 2.9.

Sharp spikes are noticed in the frequency spectrum plots. A spike is located at 12.448 GHz in Fig. 2.9a for TM mode and a spike is located at 11.142 GHz in Fig. 2.9b for TE mode. When the number of FDTD simulation steps increases, these spikes become more dominant in the frequency spectrum. The reason is that at later time, only eigen-modes exist in the computation domain whereas all other frequency components vanish. Thus, a discrete frequency spectrum can be obtained eventually. The spikes in the frequency spectrum represent the corresponding frequencies of the eigen-modes. Therefore, the frequencies 12.448 GHz and 11.142 GHz are the eigen-frequencies of the guided wave when the propagation constant k_x is 300 radian/m.

The FDTD simulation can be repeated again for another k_x value. For example, Fig. 2.10 shows the frequency spectrum at $k_x = 600$ radian/m. Two eigen-frequencies are noticed for each mode. For the TM mode, the eigen-frequencies are 21.209 GHz and 27.164 GHz. For TE mode, the eigen-frequencies are 20.017 GHz and 25.704 GHz. The FDTD results are compared with the analytical data in Table 2.1. An excellent agreement is observed and the numerical error is less than 0.15%, which demonstrates the validity of the FDTD/PBC method.

A grounded slab loaded with periodic patches [31]

The FDTD/PBC technique is also used to analyze a grounded dielectric slab loaded with periodic patches, as shown in Fig. 2.11. The thickness of the dielectric slab is 3 mm and

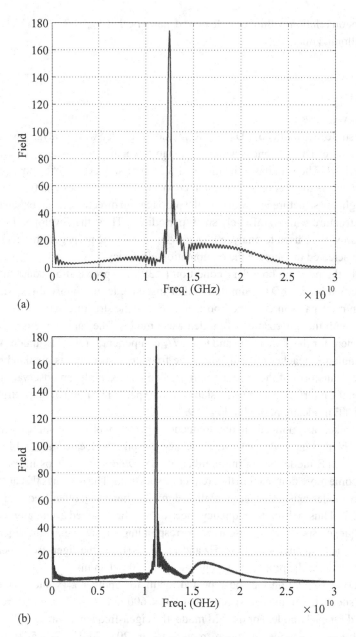

Fig. 2.9 Frequency spectrum of a dielectric slab waveguide at $k_x = 300$ radian/m: (a) TM mode and (b) TE mode.

its dielectric constant is 2.94. The width of square patch (W) is 7.5 mm and the gap width (g) between adjacent patches is 1.5 mm.

To obtain the dispersion curve for guided waves propagating along the x direction, FDTD simulations are performed 10 times, each with a specific k_x value between 0 and 250 radian/m. The results are plotted in Fig. 2.12 and each marker represents a specific surface wave mode. The same procedure is used to simulate guided waves propagating

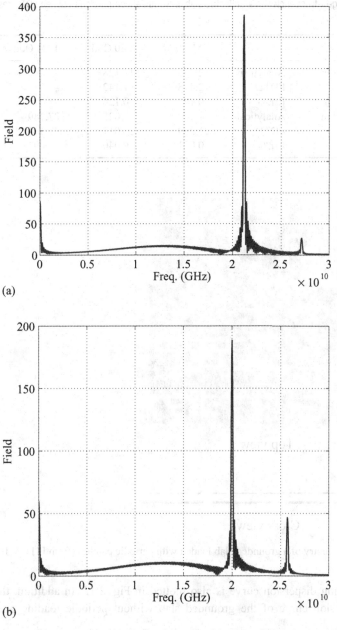

Fig. 2.10 Frequency spectrum of a dielectric slab waveguide at $k_x = 600$ radian/m: (a) TM mode and (b) TE mode.

along the diagonal direction of the patch. However, in the latter case the propagation constant k_y also needs to be considered. Besides PBC (2.39) used for boundaries along the x direction, a similar PBC needs to be used for boundaries along the y direction:

$$\vec{E}(x, y = 0, z, t) = \vec{E}(x, y = p, z, t) \cdot e^{jk_y p}$$
$$\vec{H}(x, y = 0, z, t) = \vec{H}(x, y = p, z, t) \cdot e^{jk_y p}. \quad (2.58)$$

Table 2.1 Eigen-frequencies of a dielectric slab waveguide: comparison between FDTD results and analytical data.

		TM0 Odd	TE0 Odd	TM1 Odd	TE1 Odd
$k_x = 300$ rad/m	Analytical	12.435	11.128	–	–
	FDTD	12.448	11.142	–	–
	Err.%	0.105	0.126	–	–
$k_x = 600$ rad/m	Analytical	21.222	20.025	27.189	25.740
	FDTD	21.209	20.017	27.164	25.704
	Err.%	0.061	0.040	0.092	0.140

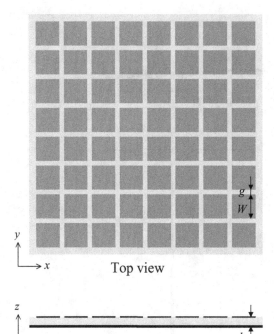

Fig. 2.11 Geometry of a grounded slab loaded with periodic patches (from [31], © IEEE, 2007).

The resulting dispersion curve is also plotted in Fig. 2.12. In addition, the light line and dispersion curve of the grounded slab without periodic loading are plotted as references.

For a thin grounded dielectric slab, only TM_0 mode exists within the interested frequency region (0–10 GHz), and the eigen-frequency at a given propagation constant (wavenumber) is slightly lower than the light line. Similar observation is found for the periodic patch loaded structure at the lower frequency region. When the frequency increases, a noticeable deviation is observed between the dispersion curves of the periodic structure and the dispersion curve of the grounded slab. At a given propagation constant, the eigen-frequency of the periodic structure is lower than that of the grounded

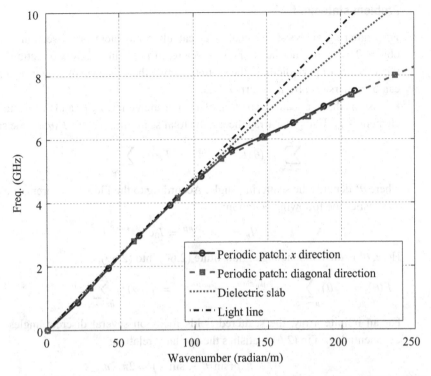

Fig. 2.12 Dispersion curves of the structure shown in Fig. 2.11 (from [31], © IEEE, 2007).

slab. This observation is similar to the effect of increasing the dielectric constant of a grounded slab waveguide.

Furthermore, it is noticed that the dispersion curve remains similar for different propagation directions, such as the x direction and the diagonal direction. The reason is that the patch size is much smaller than the wavelength at the frequency of interest. Thus, the difference in the wavenumbers between the x direction and the diagonal direction is too small to be observed.

2.4 Plane wave scattering analysis

Periodic structures also find many applications in the design of various scatterers and antenna arrays. For example, frequency selective surfaces (FSS) consisting of periodic dipoles or slots are used as radomes for large antenna systems [32]. Planar reflectarray antennas, which are built from hundreds or thousands of antenna elements arranged in a periodic lattice, are installed in spacecraft for Earth remote sensing and deep space exploration programs [33]. Recently, periodic dielectric rods and metal patterns are designed to realize electromagnetic band gap (EBG) structures and double negative metamaterials [34]. In this section, the FDTD/PBC algorithm is implemented to calculate the scattering properties of periodic structures.

2.4.1 Problem statement

For scattering analysis, the goal is to calculate the radar cross section (RCS) of an object. That is to find the strength of scattered fields in different directions. However, the presence of periodicity in the scatterer limits the existence of scattering angles. This can be understood from the array theory.

Assume that the scattered pattern of a single element is $F_e(\theta)$ and the excitation on the element is I_n. Using the array theory, the total scattering pattern $F(\theta)$ of the structure is

$$F(\theta) = \sum_{n=-\infty}^{\infty} F_e(\theta) I_n e^{jk_0 x_n \sin\theta^s} = F_e(\theta) \sum_{n=-\infty}^{\infty} I_n e^{jk_0 np \sin\theta^s}, \quad (2.59)$$

where θ^s denotes the scattering angle. According to the Floquet theorem, the excitation I_n satisfies the following equation:

$$I_n = I_0 e^{-jk_x np} = I_0 e^{-jk_0 \sin\theta^i np}. \quad (2.60)$$

Here, θ^i is the incident angle. Substitute (2.60) into (2.59),

$$F(\theta) = F_e(\theta) \sum_{n=-\infty}^{\infty} I_0 e^{-jk_0 \sin\theta^i np} e^{jk_0 np \sin\theta^s} = F_e(\theta) I_0 \sum_{n=-\infty}^{\infty} e^{jk_0 np (\sin\theta^s - \sin\theta^i)} \quad (2.61)$$

For an infinite array, the scattered fields focus on several discrete angles where the exponential term in (2.61) satisfies the in-phase relation:

$$k_0 p (\sin\theta^s - \sin\theta^i) = 2\pi \cdot m \quad (2.62)$$

where m is an integer number. Thus, the scattered angle can be calculated as below:

$$\theta^s = \sin^{-1}\left(\sin\theta^i + \frac{2\pi \cdot m}{k_0 p}\right). \quad (2.63)$$

When $m = 0$, the scattered angle θ^s is equal to the incident angle θ^i. When $m = 1$, the scattered angle θ^s is:

$$\theta^s = \sin^{-1}\left(\sin\theta^i + \frac{2\pi}{k_0 p}\right). \quad (2.64)$$

If $\sin\theta^i + \frac{2\pi}{k_0 p} \leq 1$, the corresponding scattered angle θ^s exists. Otherwise, the corresponding scattered angle θ^s does not exist. This is similar to the concept of the visible angular range in the array theory. A similar derivation applies for other integers m to determine the corresponding scattered angle θ^s.

In summary, the number of available scattered angles depends on the incident angle θ^i, the free space wavenumber k_0, and periodicity p. For example, Fig. 2.13 plots the scattering angles for a normal incident case ($\theta^i = 0$). In the low frequency region, there exists only one scattered angle that is equal to the incidence angle. As the frequency increases, the higher-order modes appear. Scattered angles for $m = \pm 1$ occur at $\pm 90°$ when the periodicity is equal to one. These two angles decrease when the frequency increases. The higher the frequency, the larger is the number of the scattered angles.

The FDTD/PBC method is used to find the field strengths at these discrete scattered angles. The corresponding reflection and transmission coefficients are calculated by the ratio of reflected field and transmitted field over the incident fields, as shown in (2.35).

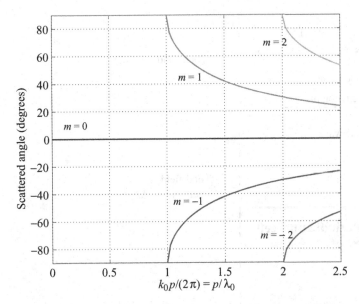

Fig. 2.13 Scattered angle of a periodic structure as a function of frequency for the normal incident case.

2.4.2 Plane wave excitation

An important issue related to the scattering analysis is the excitation of plane waves in the computational domain. The traditional approach is to use the total-field/scattered-field technique [15]. A virtual incident surface is incorporated in the computational domain, where both incident electric and magnetic fields need to be added on this interface, as shown in (2.31). This incident plane separates the computational domain into the total-field and scattered-field regions.

When this technique is applied in periodic structure analysis and the constant k_x method is used as periodic boundary conditions, a difficulty occurs regarding the incident angle. For example, if a TM^z wave impinges upon a periodic structure, the tangential incident fields H_y^{inc} and E_x^{inc} are expressed below:

$$H_y^{\text{inc}} = H_0$$
$$E_x^{\text{inc}} = -\eta H_0 \cos\theta = -\eta H_0 \sqrt{1 - (k_x/k_0)^2}, \tag{2.65}$$

where η is the wave impedance in free space. It is noticed that the E_x component depends on the incident angle. In the constant k_x method, k_x is fixed and the incident angle θ varies with frequency. Thus, it is difficult to transform the E_x in (2.65) into a time domain expression.

To solve this difficulty, the traditional total-field/scattered-field (TF/SF) technique is modified [35]: only the H_y component is added on the excitation plane for the TM^z case and the tangential E field is not incorporated. As a consequence of this one-field excitation, the plane wave is excited to propagate not only into the $z < z_0$ region but also into the $z > z_0$ region. Thus, the entire computational domain becomes the total-field

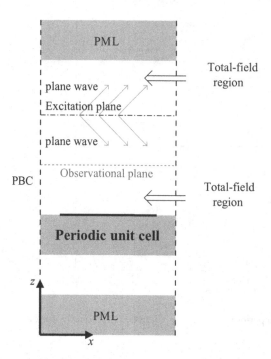

Fig. 2.14 FDTD computation model for the scattering analysis of a periodic structure.

region and there is no scattered-field region, as shown in Fig. 2.14. The scattered field is calculated by the difference of total field and incident field on an observational plane. A similar strategy applies for TE^z case: only the E_y component is added on the excitation plane. For general polarizations, it is required to break it up into TE and TM components, but both components can be excited and computed simultaneously.

On the excitation plane, the incident field adopts a modulated Gaussian waveform and the phase delay along the x direction should be inserted:

$$H_y^{\text{inc}}(x, t) = \exp\left[-\frac{(t-t_0)^2}{2\sigma_t^2}\right] \exp(j2\pi f_0 t) \exp(-jk_x x). \tag{2.66}$$

For convenience, a frequency bandwidth BW is defined below:

$$\text{BW} \times 2\pi \equiv 2 \cdot (3\sigma_f) = 2 \cdot \frac{3}{\sigma_t}. \tag{2.67}$$

The signal magnitude at $f_0 \pm (\text{BW}/2)$ is 40 dB lower than the signal level at the center frequency f_0. Since we are interested in the scattering property, the center frequency is set using the following equation:

$$f_0 = f_1 + \frac{\text{BW}}{2} = \frac{k_x c}{2\pi} + \frac{\text{BW}}{2}, \tag{2.68}$$

where f_1 is the frequency on the light line determined by the horizontal wavenumber k_x and free space wave speed c. Thus, the excitation signal strength is concentrated in the frequency range above the light line, which is in the plane wave region as shown in Fig. 2.5.

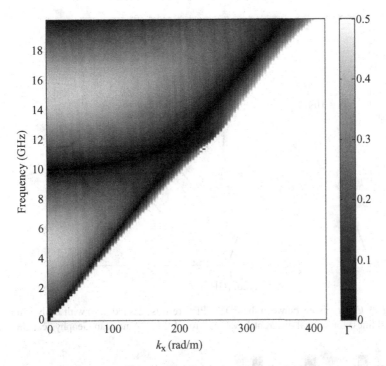

Fig. 2.15 Reflection coefficient of the dielectric slab in the k_x-frequency plane (from [35], © American Geophysical Union, 2007).

2.4.3 Examples

Reflection from a dielectric slab

The accuracy of the developed algorithm is first demonstrated through the analysis of the dielectric slab shown in Fig. 2.4. The dielectric slab can be considered as a periodic structure with infinitely small periodicity ($p \to 0$). Thus, only one scattered angle exists according to Fig. 2.13, which is equal to the incident angle. The FDTD method is used to simulate this structure and the periodic boundary condition in (2.47) is implemented.

To obtain a good resolution in the k_x-frequency plane, reflection coefficients are calculated at 100 k_x values ranging from 0 to 419 (radian/m). Thus, the FDTD simulation is repeated for 100 times and each simulation outputs the reflection coefficient on a vertical line in the k_x-frequency plane. The obtained reflection coefficient as a function of both k_x and frequency is plotted in Fig. 2.15, which is similar to the analytical result in Fig. 2.5. The half wavelength transmission region and the Brewster angle transmission region can be clearly visualized. Furthermore, Fig. 2.16 compares the analytical and FDTD computed reflection coefficients at several k_x values. The excellent agreement demonstrates the utility of the constant k_x method.

A dipole frequency selective surface (FSS)

The second example is the characterization of a frequency selective surface (FSS) consisting of dipole elements. The metallic dipole array periodically resides on a dielectric

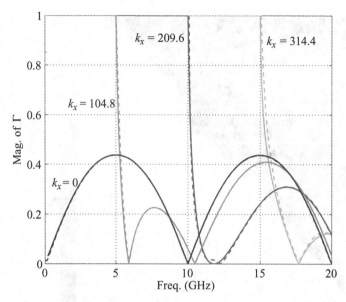

Fig. 2.16 Comparisons between the FDTD/PBC results (dashed line) with analytical results (solid lines) for several different k_x values (from [35], © American Geophysical Union, 2007).

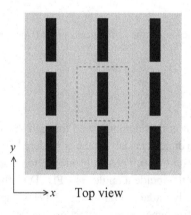

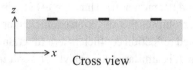

Fig. 2.17 Geometry of a frequency selective surface (FSS) consisting of dipole elements (from [35], © American Geophysical Union, 2007).

slab with a thickness h and dielectric constant ε_r. The dipole length is 12 mm and width is 3 mm. The periodicity is 15 mm in both the x and y directions. The substrate has a thickness of 6 mm and dielectric constant of 2.2. These parameters are the same as those given in [36]. A TE^z plane wave impinges on the periodic structure in the xz

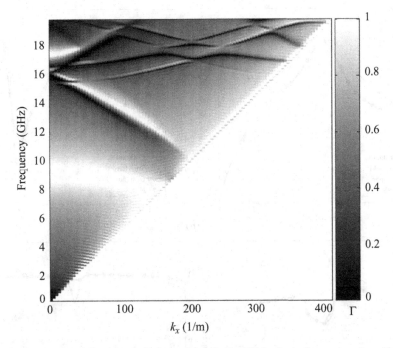

Fig. 2.18 The reflection coefficient of the dipole FSS in the k_x-frequency plane (from [35], © American Geophysical Union, 2007).

plane and the reflection coefficient for the fundamental mode ($m = 0$) in (2.63) is calculated.

The FDTD simulation is performed 100 times using different values of k_x sampled along the axis of horizontal wavenumber. The calculated reflection coefficients are plotted in Fig. 2.18. Strong reflection is represented by the white color and transmission region is denoted by the dark zone. The first reflected region occurs around 9 GHz and the frequency of total reflection slightly decreases as k_x increases, which agrees with the observation in [37]. The first transmission region starts at 16 GHz for normal incidence ($k_x = 0$). When k_x increases, both the transmission frequency and the magnitude of the transmission coefficient decrease. It is also observed from Fig. 2.18 that the second reflection region immediately follows the first transmission region. Higher order transmission and reflection regions occur at frequencies higher than 16 GHz.

The reflection coefficient versus frequency at any given incident angles can be extracted from the data on the k_x-frequency plane using an interpolation scheme. For a given incident angle θ and frequency f, the corresponding wavenumber k_x can be calculated:

$$k_x = \frac{2\pi f}{c} \cdot \sin\theta. \qquad (2.69)$$

Assume that k_{x1} and k_{x2} ($k_{x1} < k_x < k_{x2}$) are sampled horizontal wavenumbers in the FDTD simulations that are closest to k_x, the reflection coefficient Γ at k_x is calculated

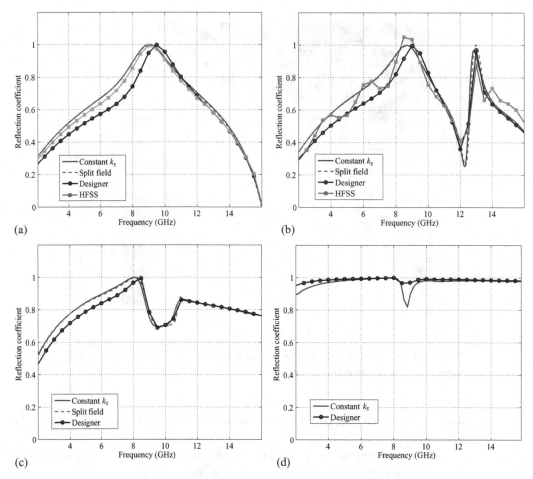

Fig. 2.19 Comparisons on reflection coefficients of the dipole FSS at several incident angles: (a) 0°, (b) 30°, (c) 60°, and (d) 85° (from [35], © American Geophysical Union, 2007).

through a linear interpolation:

$$\Gamma(f, k_x) = \frac{\Gamma(f, k_{x1}) \cdot (k_{x2} - k_x) + \Gamma(f, k_{x2}) \cdot (k_x - k_{x1})}{k_{x2} - k_{x1}} \quad (2.70)$$

For example, the reflection coefficients for the incident angles of 0°, 30°, 60°, and 85° are extracted and plotted in Fig. 2.19.

To verify the accuracy of the FDTD results, the same structure is analyzed using three different programs. One is the split-field FDTD program developed at the University of California, Los Angeles [38]. The other two are the Ansoft HFSS program and the Ansoft Designer program. The HFSS program is based on the Finite Element Method (FEM), and the Designer program is based on the Method of Moments (MoM). All three programs compute the reflection coefficient at given incident angles. Good agreements between different methods are observed, especially at small incident angles. As the incident angle increases, the HFSS results become unstable. When the incident angle is close to 90°, the split-field FDTD method cannot give a converged result because of the prohibitively long simulation time.

It is worthwhile to emphasize an advantage of the constant k_x method in this example. When the field transformation methods such as the split-field technique are used to calculate the oblique incidence, the time step size (Δt) needs to be decreased as the incident angle (θ) increases. To satisfy the stability constrains, the time step size and incident angle for square cells and free space should follow the inequality below,

$$\frac{c\Delta t}{\Delta x} \leq \frac{1-\sin\theta}{\sqrt{3}}. \tag{2.71}$$

Δx is the size of the FDTD grid cell. The time step size approaches zero when the incident angle reaches 90°. For example, when Δx is 1 mm in the FSS analysis, $\Delta t < 1.925 \times 10^{-12}$ (seconds) is required for normal incidence. For a large incident angle such as 85°, $\Delta t < 7.323 \times 10^{-15}$ (seconds) is necessary. Therefore, the simulation time becomes prohibitively long for the latter case.

In contrast, the constant k_x method uses a standard Yee's scheme to update the E and H fields. The stability condition remains unchanged regardless of the horizontal wavenumber (k_x) or incident angle, as expressed below:

$$\frac{c\Delta t}{\Delta x} \leq \frac{1}{\sqrt{3}}. \tag{2.72}$$

Thus, a constant $\Delta t = 1.667 \times 10^{-12} < 1.925 \times 10^{-12}$ (seconds) is used in FDTD simulation for all different k_x values. Therefore, this approach has a good computational efficiency. Especially for large incident angles, the time step size of the constant k_x method is three orders greater than the split-field method.

2.5 A unified approach: hybrid FDTD/ARMA method

The periodic boundary condition based on the constant k_x method is applicable for both the guided wave analysis and scattering calculation, as demonstrated in the previous two sections. Thus, it provides an opportunity to combine the guided wave and scattering analysis in a single FDTD simulation. This section introduces the unified FDTD approach [39], including the concept, the oscillation problem, and the auto-regressive moving average (ARMA) estimator [40] in time domain data processing. The accuracy of the unified approach is demonstrated by several examples.

2.5.1 A unified approach for guided wave and scattering analysis

Surface wave region and plane wave region

To understand the unified approach, let's revisit the definitions of the surface wave region and the plane wave region. For an electromagnetic wave at a given frequency f, the free space wavenumber is calculated as below:

$$k_0 = \omega\sqrt{\varepsilon_0\mu_0} = 2\pi f\sqrt{\varepsilon_0\mu_0}. \tag{2.73}$$

The propagation vector $\vec{k}$ can be decomposed in the rectangular coordinate:

$$\begin{aligned}\vec{k} &= \hat{x}k_x + \hat{y}k_y + \hat{z}k_z \\ \left|\vec{k}\right|^2 &= k_0^2 = \omega^2\varepsilon_0\mu_0 = k_x^2 + k_y^2 + k_z^2.\end{aligned} \tag{2.74}$$

For a given combination of horizontal wavenumbers (k_x, k_y), if the frequency ω is small and the following equation is satisfied,

$$k_0^2 = \omega^2\varepsilon_0\mu_0 < k_x^2 + k_y^2, \tag{2.75}$$

then the value of k_z must be a purely imaginary number in order to satisfy (2.74). As a result, the field decays exponentially in the z direction. The electromagnetic wave is a surface wave propagating horizontally. If the frequency ω is large and the following equation is satisfied,

$$k_0^2 = \omega^2\varepsilon_0\mu_0 > k_x^2 + k_y^2, \tag{2.76}$$

then the value of k_z is a real number. The field also propagates along the z direction and the electromagnetic wave is a plane wave. When the frequency ω satisfies the following equation:

$$k_0^2 = \omega^2\varepsilon_0\mu_0 = k_x^2 + k_y^2, \tag{2.77}$$

the wavenumber $k_z = 0$. The field has no variation in the z direction. The frequency is located on the so-called light line, which separates the surface wave region and the plane wave region. In summary, the FDTD simulation at a given combination of horizontal wavenumbers (k_x, k_y) could cover both surface wave region and plane wave region if the wideband excitation signal includes both frequency components that satisfy (2.75) and (2.76).

Surface impedance and wave impedance

Next, an appropriate characteristic parameter needs to be adopted to describe both the guided wave feature and scattering property. In the guided wave analysis in Section 2.3, the eigen-frequencies and eigen-modes are of interest. In the scattering analysis in Section 2.4, the reflection coefficient is determined. Here, we use the concept of wave impedance. As shown in Fig. 2.14, an observation plane is set in the computational domain and the tangential electric and magnetic fields are collected. The total E and H fields on the surface are extracted as following with the phase delay being considered:

$$\begin{aligned}E_s &= \int_S E(x,y)e^{j(k_x x + k_y y)}ds \\ H_s &= \int_S H(x,y)e^{j(k_x x + k_y y)}ds\end{aligned} \tag{2.78}$$

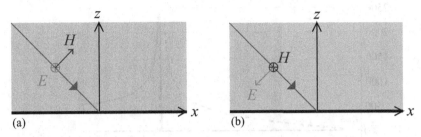

Fig. 2.20 (a) TE and (b) TM incidences on a grounded dielectric slab.

Then, the impedance at the observation plane is calculated by taking the ratio of E and H field:

$$Z_s = E_s/H_s. \tag{2.79}$$

This impedance definition applies to both the surface wave region and the plane wave region. From the surface impedance, the reflection coefficient and the eigen-frequency can be easily derived. Similar to the transmission line theory, the reflection coefficient can be calculated as following:

$$\Gamma = \frac{Z_s - Z_w}{Z_s + Z_w}, \tag{2.80}$$

where Z_w is the tangential wave impedance of the incident wave in the free space. For TM and TE plane waves, the wave impedances are:

$$Z_w^{TM} = \frac{k_z}{k_0}\eta_0, \quad Z_w^{TE} = \frac{k_0}{k_z}\eta_0. \tag{2.81}$$

Note that in the surface wave region, k_z is an imaginary number and the Z_w is also imaginary. Both Z_s and Z_w are functions of frequency. When $Z_s + Z_w = 0$, the reflection coefficient in (2.80) approaches to infinity. It means that the wave can exist without an excitation, which is an eigen-mode in a waveguide. Thus, the corresponding frequency is the eigen-frequency for that mode.

To illustrate the concepts above, a grounded dielectric slab is used as an example. Figure 2.20 shows the TE and TM incidences on a grounded slab. The surface impedances can be derived analytically:

$$Z_s^{TE} = \frac{E_y}{H_x} = \frac{j\omega\mu}{k_z}\tan(k_z h)$$
$$Z_s^{TM} = \frac{-E_x}{H_y} = \frac{jk_z}{\omega\varepsilon}\tan(k_z h) \tag{2.82}$$

It is worthwhile to point out that the surface impedances always have imaginary values, no matter in the plane wave region or surface wave region. The reason is that there is no loss in this structure and the real part of the surface impedance must be zero.

Figure 2.21 shows the surface impedances and wave impedances for the TE case and TM case. The thickness of the grounded slab is 4 mm and the dielectric constant is 4. The value of k_x is set to 222.14 radian/m. As discussed, the real parts of the surface impedances

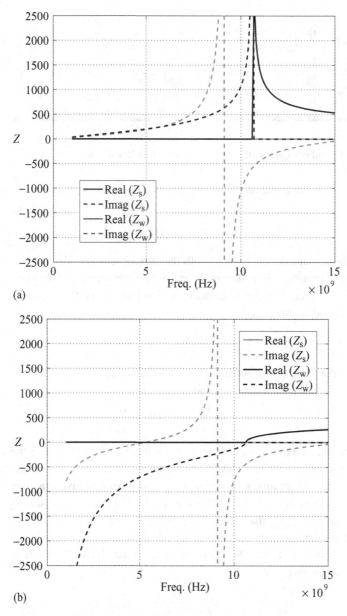

Fig. 2.21 Surface impedances of the grounded slab and wave impedance of an incident plane wave: (a) TE case and (b) TM case.

are always zero where the imaginary parts vary with frequency. The wave impedance is imaginary when the frequency is smaller than 10.61 GHz, the corresponding light line frequency to the k_x. In contrast, the wave impedance is real when the frequency is larger than 10.61 GHz. In the plane wave region, since the surface impedance is imaginary and wave impedance is real, the reflection coefficient always has a unity magnitude, which means the energy is totally reflected. This is reasonable because of the ground plane in this structure. In the surface wave region, both impedances are imaginary and when their values are opposite, the eigen-frequencies can be identified.

2.5.2 ARMA estimator

With the concepts discussed above, the FDTD method is used in the analysis with the constant k_x method as the periodic boundary condition. The same excitation technique as Section 2.4.2 is used but the signal bandwidth is selected to be wide enough to cover both the surface wave region and the plane wave region. The surface impedance is calculated using (2.78) and (2.79).

As an example, Fig. 2.22 shows the time domain data of TE incident waves on a grounded dielectric slab with different k_x values. Resonant behavior is observed in the time domain data. At the eigen-frequencies, the surface waves are guided along the horizontal directions, the energy exiting the computation domain at boundary $x = p$ will re-enter the domain at boundary $x = 0$ due to the implementation of the periodic boundary condition. Consequently, the time domain data will not decay to zero and a resonance behavior is formed. More severely, the resonance behavior may cause instability in the time domain computation. It is observed in Fig. 2.22a that a small numerical error could be amplified during the resonant procedure. As a consequence, accurate frequency domain features cannot be obtained through the usual Fourier transformation process.

In practice, resonance in the time domain data is an important phenomenon in FDTD simulations of various electromagnetic objects, such as microstrip antennas, microwave filters, and frequency selective surfaces. Several signal processing techniques have been utilized to process the time domain data so that the desired characteristic parameters can be accurately obtained. Typical techniques include Prony's method, system identification (SI) method, Pade approximation, and pencil function [41–46]. In [47–48], the ARMA estimator, a popular technique in the signal processing area, is used in the FDTD method for determinant problems such as antenna impedance and reflection coefficient. Here, the ARMA estimator is used in the unified FDTD simulation of both determinant and eigen-value problems.

As a starting point, the analyzed structure is considered as a linear system with the following transfer function:

$$H(z) = \frac{a_0 + a_1 Z^{-1} + a_2 Z^{-2} + \cdots + a_q Z^{-q}}{1 + b_1 Z^{-1} + b_2 Z^{-2} + \cdots + b_p Z^{-p}}. \tag{2.83}$$

The transfer function can also be expressed in the time domain as follows:

$$y(n) = -\sum_{i=1}^{p} b_i y(n-i) + \sum_{j=0}^{q} a_j x(n-j) \tag{2.84}$$

where $x(n)$ and $y(n)$ are input and output signals of the linear system. b_i and a_j are unknown coefficients to be determined.

As the FDTD simulation is performed, the excitation signal is used as the input signal $x(n)$ and the total field E_s or H_s calculated in (2.78) is used as the output signal $y(n)$.

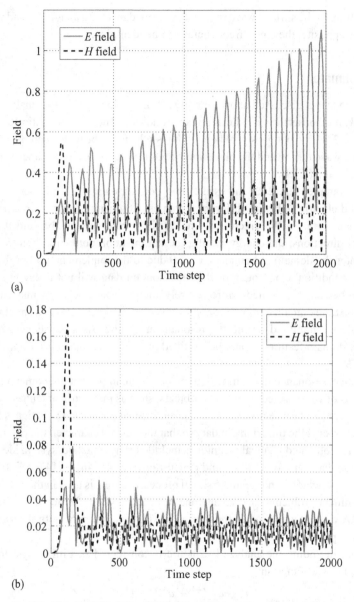

Fig. 2.22 Time domain data of TE incident waves on a grounded dielectric slab: (a) $k_x = 300$ rad/m, (b) $k_x = 600$ rad/m.

If the time step size in the FDTD simulation is very small in the frequency band of interest, $x(n)$ and $y(n)$ could be sampled from the excitation signal and the field data. The sample rate is determined by the frequency and FDTD time step size. When the FDTD simulation proceeds to obtain the Nth samples, a linear equation can be established from (2.84):

$$[Y]_{N\times 1} = [D]_{N\times (p+q+1)}[C]_{(p+q+1)\times 1} \qquad (2.85)$$

where [Y] is the output vector, [C] is the coefficient vector, and [D] is a matrix generated by the input and output data, as expressed below:

$$[Y] = [y(1), y(2), \ldots, y(N)]' \in R^{N \times 1} \quad (2.86)$$

$$[C] = [b_1, \ldots, b_p, a_0, \ldots, a_q] \in R^{(p+q+1) \times 1} \quad (2.87)$$

$$[D] = \begin{bmatrix} 0 & 0 & \cdots & 0 & x(1) & 0 & \cdots & 0 \\ -y(1) & 0 & \cdots & 0 & x(2) & x(1) & \cdots & 0 \\ \vdots & \ddots & \ddots & 0 & \vdots & \ddots & \ddots & 0 \\ y(p) & y(p-1) & \cdots & -y(1) & x(p+1) & x(p) & \cdots & x(p+1-q) \\ -y(p+1) & -y(p) & \cdots & -y(2) & x(p+2) & x(p+1) & \cdots & x(p+2-q) \\ \vdots & & \ddots & \vdots & \vdots & & \ddots & \vdots \\ -y(N-1) & -y(N-2) & \cdots & -y(N-p) & x(N) & x(N-1) & \cdots & x(N-q) \end{bmatrix} \in R^{N \times (p+q+1)} \quad (2.88)$$

[C] is the unknown coefficient to be determined and the number of unknowns is $p + q + 1$. When the sampling number N is larger than $p + q + 1$, the coefficients can be determined using the least mean square error estimation:

$$[C] = ([D]^T[D])^{-1}[D]^T[V]. \quad (2.89)$$

Once [C] is determined, the frequency domain transfer function $H(j\omega)$ can be obtained from (2.83) by replacing Z with $\exp(j\omega)$. Note that the frequency ω here is a normalized frequency with respect to the sampled time step.

The ARMA estimator exhibits several advantages in FDTD applications. First, the frequency response can be obtained accurately. This will be demonstrated in the following examples. Second, it improves the FDTD computation efficiency for resonant problems. In general, the FDTD simulation needs to be performed for a long time in order to reach a steady state. However, the ARMA estimator uses only the early time domain data for the process. Thus, it significantly reduces the number of time steps required in the FDTD simulation. In addition, the eigen-frequencies can be directly obtained from the transfer function in (2.83). When the transfer function has poles located on the unit circle, the corresponding frequencies are the eigen-frequencies for the system.

2.5.3 Examples

Unified analysis of a grounded dielectric slab

The validity of the unified FDTD method is first demonstrated in the analysis of a grounded dielectric slab. The thickness of the slab is 6.35 mm and the dielectric constant is 2.65. A perfect electric conductor is positioned on the back of the slab. The structure is simulated using the hybrid FDTD/ARMA method and the results are compared with the analytical data. An ARMA estimator with $p = q = 40$ is used to solve the difficulties from the resonance phenomenon.

Since the surface is lossless, the computed surface impedance is purely imaginary, as derived in (2.82). Figure 2.23a shows the imaginary part of surface impedance for TE waves and Fig. 2.23b for TM waves. The impedance obtained covers both surface wave region and plane wave region. For example, when $k_x = 300$ rad/m, the surface

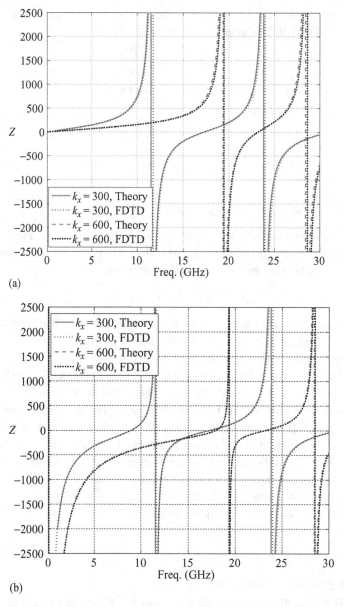

Fig. 2.23 Surface impedance of a grounded dielectric slab: (a) TE case, (b) TM case.

wave region is from 0–14.3 GHz and the plane wave region starts from 14.3 GHz. The eigen-frequencies of surface wave modes are also identified from the ARMA transfer functions. The results are listed in Table 2.2. Excellent agreement between the FDTD data and the theoretical data is observed.

Corrugated soft/hard surface

A corrugated surface can be realized by adding metal walls to a grounded dielectric slab, as shown in Fig. 2.24. This surface operates as a soft surface for an electromagnetic

Table 2.2 Eigen-frequencies of surface wave modes

	$k_x = 300$ rad/m TE mode	$k_x = 300$ rad/m TM mode	$k_x = 600$ rad/m TE mode	$k_x = 600$ rad/m TM mode
Theory (GHz)	12.87	10.61	21.12	19.03
FDTD (GHz)	12.75	10.57	21.11	19.01
Error (%)	0.93	0.38	0.05	0.11

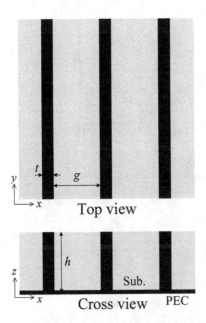

Fig. 2.24 Geometry of a corrugated soft/hard surface: $g = 4$ mm, $t = 1$ mm, $h = 5$ mm, $\varepsilon_r = 4.0$ (from [39], © IEEE, 2005).

wave propagating along the x direction, and as a hard surface for an electromagnetic wave propagating along the y direction. This artificial structure has been used in waveguide designs and corrugated horn antennas.

The TM and TE impedances of the surface are calculated for several k_x and k_y values. When the wave propagates along the x direction, k_x is set to 0, 50π radian/m, and 100π radian/m while k_y is set to 0. As shown in Fig. 2.25, a low TE impedance and a high TM impedance are obtained, which represents a soft operation that stops the wave propagation. It is interesting to notice that both TM and TE impedances are almost independent of the wavenumber of the incident wave. Thus, the soft operation is independent of the incident angle.

When the wave propagates along the y direction, k_y is set to 0, 50π radian/m, and 100π radian/m while k_x is set to 0. As shown in Fig. 2.26, a high TE impedance and a low TM impedance are observed, which indicates a hard operation that allows the wave to propagate. For y-propagating waves, the wave impedances are sensitive to the wavenumbers. As the wavenumber increases, the surface impedances move along the high frequency direction.

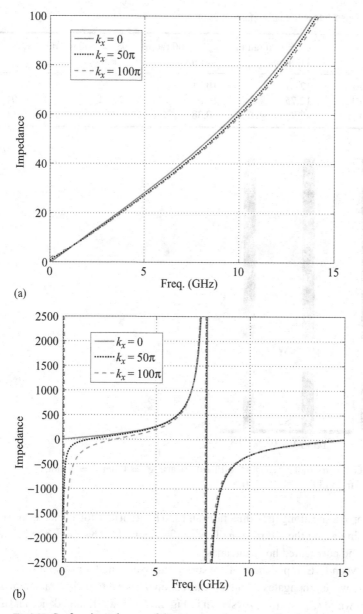

Fig. 2.25 Surface impedances of the corrugated structure for waves propagating along the x direction. (a) A low TE impedance and (b) a high TM impedance are obtained for a soft operation.

2.6 Projects

Project 2.1 Write a FDTD program with the implementation of the periodic boundary condition (PBC) in (2.47). Use your code to analyze the dielectric slab waveguide and dipole frequency selective surface discussed in this chapter.

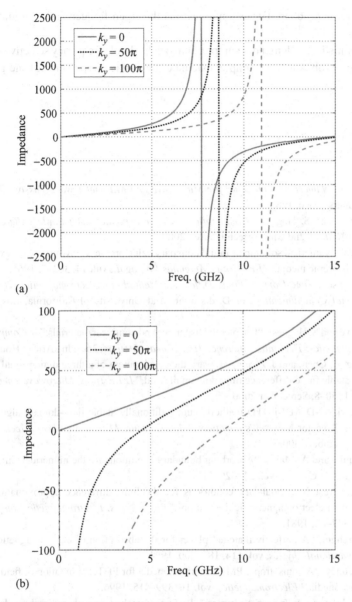

Fig. 2.26 Surface impedances of the corrugated structure for waves propagating along the y direction. (a) A high TE impedance and (b) a low TM impedance are obtained for a hard operation.

Project 2.2 Derive the surface impedance formula of a grounded dielectric slab and compare your results with (2.82). Plot surface impedance curves for different k_x values, as illustrated in Fig. 2.23.

Project 2.3 Write a Matlab program based on the ARMA estimator formulas from (2.83) to (2.89). Use your program to process the FDTD data obtained from Project 2.1. Three types of problems should be tested:

(a) Eigen-value problem. For example, extract the eigen-frequencies of a guided wave structure.
(b) Deterministic problem. For example, analyze the dipole frequency selective surfaces.
(c) A hybrid problem. For example, calculate the surface impedance of the grounded slab.

References

1. T. Itoh, ed. *Numerical Techniques for Microwave and Millimeter-wave Passive Structures*, Wiley-Interscience, 1989.
2. A. Taflove and S. Hagness, *Computational Electrodynamics: The Finite Difference Time Domain Method*, 2nd edn., Artech House, 2000.
3. K. S. Yee, "Numerical solution of initial boundary value problems involving Maxwell's equations in isotropic media," *IEEE Trans. Antennas Propagat.*, **vol. 14**, 302–7, 1966.
4. M. A. Jensen, *Time-Domain Finite-Difference Methods in Electromagnetics: Application to Personal Communication*, Ph.D. dissertation at University of California, Los Angeles, 1994.
5. A. Taflove and S. Hagness, "Chapter 9: Dispersive and nonlinear materials," in *Computational Electrodynamics: The Finite Difference Time Domain Method*, 2nd edn., Artech House, 2000.
6. F. Zheng, Z. Chen, and J. Zhang, "Toward the development of a three dimensional unconditionally stable finite-difference time-domain method," *IEEE Trans. Microwave Theory Tech.*, **vol. 48**, 1550–8, September 2000.
7. T. NaMiki, "3-D ADI-FDTD method – unconditionally stable time-domain algorithm for solving full vector Maxwell's equations," *IEEE Trans. Microwave Theory Tech.*, **vol. 48**, 1743–8, October 2000.
8. B. Engquist and A. Majda, "Absorbing boundary conditions for the numerical simulation of waves," *Math. Comput*, **vol. 31**, 629–51, 1977.
9. G. Mur, "Absorbing boundary conditions for the finite-difference approximation of the time-domain electromagnetic field equations," *IEEE Trans. Electromagnetic Compatibility*, **vol. 23**, 377–82, 1981.
10. J. P. Berenger, "A perfectly matched player for the absorption of electromagnetic waves," *J. Computational Physics*, **vol. 114**, 185–200, 1994.
11. S. D. Gedney, "An anisotropic PML absorbing media for FDTD simulation of fields in lossy dispersive media," *Electromagnetics*, **vol. 16**, 399–415, 1996.
12. S. D. Gedney and A. Taflove, "Chapter 7: Perfectly matched layer absorbing boundary conditions," in *Computational Electrodynamics: The Finite Difference Time Domain Method*, 2nd edn., A. Taflove and S. Hagness, Artech House, 2000.
13. M. A. Jensen and Y. Rahmat-Samii, "Performance analysis of antennas for hand-held transceiver using FDTD," *IEEE Trans. Antennas Propagat.*, **vol. 42**, 1106–13, 1994.
14. A. Z. Elsherbeni and Y. Rahmat-Samii, *FDTD Analysis of Printed Microstrip Antennas for Personal Communication*, Technical Report, Department of Electrical Engineering, University of California, Los Angeles, December 1996.
15. A. Taflove and S. Hagness, "Chapter 5: Incident wave source conditions," in *Computational Electrodynamics: The Finite Difference Time Domain Method*, 2nd edn., Artech House, 2000.

16. J. Maloney and M. P. Kesler, "Chapter 13: Analysis of periodic structures," in *Computational Electrodynamics: The Finite Difference Time Domain Method*, 2nd edn., A. Taflove and S. Hagness, Artech House, 2000.
17. L. Brillouin, *Wave Propagation in Periodic Structures*, 2nd edn., Dover Publications, 2003.
18. S. Xiao, R. Vahldieck, and H. Jin, "Full-wave analysis of guided wave structures using a novel 2-D FDTD," *IEEE Microw. Guided Wave Lett.*, **vol. 2, no. 5**, 165–7, 1992.
19. A. C. Cangellaris, M. Gribbons, and G. Sohos, "A hybrid spectral/FDTD method for electromagnetic analysis of guided waves in periodic structures," *IEEE Microw. Guided Wave Lett.*, **vol. 3, no. 10**, 375–7, 1992.
20. H. S. Chan, S. H. Lou, L. Tsang, and J. A. Kong, "Electromagnetic scattering of waves by random rough surface: a finite difference time domain approach," *Microwave Optical Tech. Lett.*, **vol. 4**, 355–9, 1991.
21. P. Harms, R. Mittra, and W. Ko, "Implementation of the periodic boundary condition in the finite-difference time-domain algorithm for FSS structures," *IEEE Trans. Antennas and Propagation*, **vol. 42**, 1317–24, 1994.
22. J. A. Roden, S. D. Gedney, M. P. Kesler, J. G. Maloney, and P. H. Harms, "Time-domain analysis of periodic structures at oblique incidence: orthogonal and nonorthogonal FDTD implementations," *IEEE Trans. Microwave Theory and Techniques*, **vol. 46, no. 4**, 420–7, 1998.
23. M. E. Veysoglu, R. T. Shin, and J. A. Kong, "A finite-difference time-domain analysis of wave scattering from periodic surfaces: oblique incidence case," *J. Electromagnetic Waves and Applications*, **vol. 7**, 1595–607, 1993.
24. J. Ren, O. P. Gandhi, L. R. Walker, J. Fraschilla, and C. R. Boerman, "Floquet-based FDTD analysis of two-dimensional phased array antennas," *IEEE Microwave and Guided Wave Lett.*, **vol. 4**, 109–12, 1994.
25. H. Holter and H. Steyskal, "Infinite phased-array analysis using FDTD periodic boundary conditions – pulse scanning in oblique directions," *IEEE Trans. Antennas and Propagation*, **vol. 47**, 1508–14, 1999.
26. J. R. Marek and J. MacGillivary, "A method for reducing run-times of out-of-core FDTD problems," *Proc. 9th Annual Review of Progress in Applied Computational Electromagnetics*, Montery, CA, pp. 344–51, March 1993.
27. A. Aminian and Y. Rahmat-Samii, "Spectral FDTD: a novel computational technique for the analysis of periodic structures," *IEEE APS Int. Symp. Dis.*, vol. 3, pp. 3139–42, June 2004.
28. F. Yang, J. Chen, R. Qiang, and A. Elsherbeni, "FDTD analysis of periodic structures at arbitrary incidence angles: a simple and efficient implementation of the periodic boundary conditions," *IEEE APS Int. Symp. Dis.*, vol. 3, pp. 2715–18, 2006.
29. R. E. Collin, *Field Theory of Guided Waves*, 2nd edn., Wiley-IEEE Express, 1990.
30. K. Zhang and D. Li, *Electromagnetic Theory for Microwaves and Optoelectronics*, 2nd edn., Publishing House of Electronics Industry, 2001.
31. A. Al-Zoubi, F. Yang, and A. Kishk, "A low profile dual band surface wave antenna with a monopole like pattern," *IEEE Trans. Antennas Propagat.*, **vol. 55, no. 12**, 3404–12, 2007.
32. B. A. Munk, *Frequency Selective Surface*, John Wiley & Sons, Inc., 2000.
33. J. Huang, "The development of inflatable array antennas," *IEEE Antennas Propagat. Mag.*, **vol. 43, no. 4**, 44–50, 2001.
34. N. Engheta and R. Ziolkowski, *Metamaterials: Physics and Engineering Explorations*, John Wiley & Sons Inc., 2006.

35. F. Yang, J. Chen, Q. Rui, and A. Elsherbeni, "A simple and efficient FDTD/PBC algorithm for periodic structure analysis," *Radio Sci.*, **vol. 42, no. 4**, RS4004, July 2007.
36. J. Gianvittorio, J. Romeu, S. Blanch, and Y. Rahmat-Samii, "Self-similar prefractal frequency selective surfaces for multiband and dual-polarized applications," *IEEE Trans. Antennas and Propagation*, **vol. 51, no. 11**, 3088–96, 2003.
37. E. L. Pelton and B. A. Munk, "Scattering from periodic arrays of crossed dipoles," *IEEE Trans. Antennas and Propagation*, **vol. 27, no. 3**, 323–30, 1979.
38. H. Mosallaei and Y. Rahmat-Samii, "Periodic bandgap and effective dielectric materials in electromagnetics: characterization and applications in nanocavities and waveguides," *IEEE Trans. Antennas and Propagation*, **vol. 51, no. 3**, 549–63, 2003.
39. A. Aminian, F. Yang, and Y. Rahmat-Samii, "Bandwidth determination for soft and hard ground planes by spectral FDTD: a unified approach in visible and surface wave regions," *IEEE Trans. Antennas Propagat.*, **vol. 53, no. 1**, 18–28, 2005.
40. F. Yang, A. Elsherbeni, and J. Chen, "A hybrid spectral-FDTD/ARMA method for periodic structure analysis," *IEEE APS Int. Symp. Dis.*, Hawaii, June 2007.
41. Y. Hua and T. K. Sarkar, "Matrix pencil method for estimating parameters of exponentially damped/undamped sinusoids in noise," *IEEE Trans. Accoustics Speech and Signal Processing*, **vol. 38, no. 5**, 1990.
42. W. Ko and R. Mittra, "A combination of FD-TD and Prony's methods for analyzing microwave integrated circuits," *IEEE Trans. Microwave Theory Tech.*, **vol. 39**, 2176–81, 1991.
43. Z. Bi, Y. Shen, K. Wu, and J. Litva, "Fast FDTD analysis of resonators using digital filtering and spectrum estimation," *IEEE Trans. Microwave Theory Tech.*, **vol. 40**, 1611–19, 1992.
44. W. Kuempel and L. Wolff, "Digital signal processing of time domain field simulation results using the system identification method," *IEEE MTT-S Int. Symp. Dig.*, vol. 2, pp. 793–6, June 1992.
45. B. Housmand, T. W. Huang, and T. Itoh, "Microwave structure characterization by a combination of FDTD and system identification methods," *IEEE Microwave Guided Wave Lett.*, **vol. 3**, 262–4, August 1993.
46. J. Chen, C. Wu, T. Lo, K.-L. Wu, and J. Litva, "Using linear and non-linear predictors to improve the computational efficiency of the FD-TD algorithm," *IEEE Trans. Microwave Theory Tech.*, **vol. 42**, 1992–7, September 1994.
47. A. K. Shaw and K. Naishadham, "ARMA-based time-signature estimator for analyzing resonant structures by the FDTD method," *IEEE Trans. Antennas Propagat.*, **vol. 49**, 327–39, 2001.
48. F. Yang and Y. Rahmat-Samii, "Microstrip antenna analysis using fast FDTD methods: a comparison of Prony and ARMA techniques," *Proceedings of 3rd International Conference on Microwave and Millimeter Wave Technology*, 661–4, Beijing, August 17–19, 2002.

3 EBG characterizations and classifications

Over the last decade, diversified and novel electromagnetic band gap (EBG) structures have appeared in the literature. They exhibit interesting electromagnetic properties, which are not readily available in natural materials. In this chapter, we illustrate these interesting properties of EBG structures. A classification of various EBG structures is also provided.

3.1 Resonant circuit models for EBG structures

To more readily understand the operation mechanism of EBG structures, some circuit models have been proposed. Let's start with a simple two-dimensional planar electromagnetic band gap (EBG) structure, as shown in Fig. 3.1. This structure was originally proposed in [1]. The EBG structure consists of four parts: a metal ground plane, a dielectric substrate, periodic metal patches on top of the substrate, and vertical vias connecting the patches to the ground plane. The geometry is similar to the shape of a mushroom.

3.1.1 Effective medium model with lumped LC elements

The parameters of the EBG structure are labeled in Fig. 3.2a as patch width W, gap width g, substrate thickness h, dielectric constant ε_r, and vias radius r. When the periodicity $(W + g)$ is small compared to the operating wavelength, the operation mechanism of this EBG structure can be explained using an effective medium model with equivalent lumped LC elements, as shown in Fig. 3.2b [2]. The capacitor results from the gap between the patches and the inductor results from the current along adjacent patches. The impedance of a parallel resonant LC circuit is given by:

$$Z = \frac{j\omega L}{1 - \omega^2 LC}. \qquad (3.1)$$

The resonance frequency of the circuit is calculated as following:

$$\omega_0 = \frac{1}{\sqrt{LC}}. \qquad (3.2)$$

At low frequencies, the impedance is inductive and supports TM surface waves. It becomes capacitive at high frequencies and TE surface waves are supported. Near the resonance frequency ω_0, high impedance is obtained and the EBG does not support any surface waves, resulting in a frequency band gap. The high surface impedance also

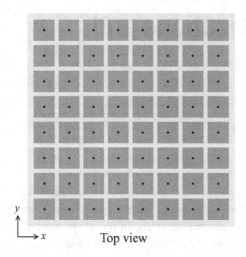

Fig. 3.1 Geometry of a mushroom-like electromagnetic band gap (EBG) structure.

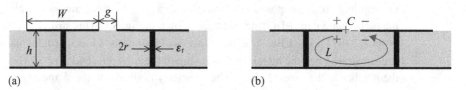

Fig. 3.2 LC model for the mushroom-like EBG structure: (a) EBG parameters and (b) LC model.

ensures that a plane wave will be reflected without the phase reversal that occurs on a perfect electric conductor (PEC).

The value of the capacitor is given by the fringing capacitance between neighboring coplanar metal plates. This can be derived using conformal mapping, a common technique for determining two-dimensional electrostatic field distributions. The derivation starts with a pair of semi-infinite plates separated by a gap and then truncates them with a finite patch size. Finally, the edge capacitance for the narrow gap situation is given by the following equation [2]:

$$C = \frac{W\varepsilon_0(1+\varepsilon_r)}{\pi} \cosh^{-1}\left(\frac{W+g}{g}\right). \tag{3.3}$$

The value of the inductor is derived from the current loop in Fig. 3.2b, consisting of the vias and metal sheets. For a solenoid current, the magnetic field can be calculated using Ampere's law. The equivalent inductor is then computed from the stored magnetic field energy and the excitation current. After a simple derivation, the inductance is expressed as below [2], which depends only on the thickness of the structure and the permeability:

$$L = \mu h. \tag{3.4}$$

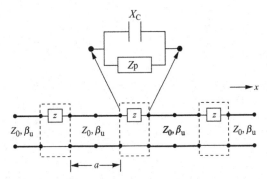

Fig. 3.3 Equivalent transmission line model for surface waves (from [3], © Wiley Inter-Science, 2007).

Substituting (3.3) and (3.4) into (3.1) and (3.2), the surface impedance and resonant frequency can be computed. Other characteristic parameters, such as the reflection phase, can also be derived accordingly. This LC model is easy to understand, but the static field approximations limit its accuracy.

3.1.2 Transmission line model for surface waves

Periodic transmission line method is another analytical model proposed to characterize EBG structures [3]. For a 2-D EBG structure such as in Fig. 3.1, the impedance of each section is calculated using transmission line theory, and the whole structure is then cascaded together using the theory of periodic circuits.

Figure 3.3 depicts a transmission line model of EBG structures. Between two nodes of the periodic structure, there are two contributions (Z_P and X_C) to the total impedance. The centrally located shorting pin in Fig. 3.1 provides the inductive loading X_l of the resonator with the inductance value equal to [3]:

$$L = 2 \times 10^{-7} h \left[\ln\left(\frac{2h}{r}\right) + 0.5\left(\frac{2r}{h}\right) - 0.75 \right]. \qquad (3.5)$$

The impedance Z_P of each resonator section is calculated using the well-known transmission line formula:

$$Z_p = Z_0 \frac{Z_1 + jZ_0 \tan(\beta_u l)}{Z_0 + jZ_1 \tan(\beta_u l)}, \qquad (3.6)$$

where Z_0 is the characteristics impedance and β_u is the phase constant of the transmission line. The transmission line is approximated by a microstrip line with the same line width and substrate properties as the mushroom-like EBG structure. The coupling capacitor X_C between the resonators is also calculated using (3.3).

Once the impedance of the resonator (Z_P) and the coupling capacitor (X_C) are known, the EBG structure can be treated as a transmission line periodically loaded with a lumped impedance Z. The value of Z is calculated from the parallel connection of Z_P and X_C.

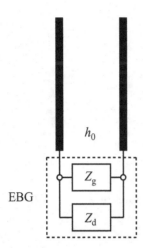

Fig. 3.4 Equivalent transmission line model for plane wave incidences.

The phase constant β for an infinite lossless periodic structure with a periodicity of a is [4]

$$\cos(\beta a) = \cos(\beta_u a) + j\frac{Z}{2Z_0}\sin(\beta_u a). \tag{3.7}$$

Note that Z is imaginary for a lossless resonator. At different frequencies, the corresponding propagation constant is calculated using (3.7) and hence a dispersion diagram can be obtained.

This method was used to analyze a mushroom-like EBG structure in [1] and a uni-planar EBG structure in [5], and satisfactory results were obtained [3]. A challenge in this method is how to accurately obtain the equivalent Z_P and X_C values for general EBG structures with arbitrary geometries. In addition, the proposed method is only used to analyze surface waves that propagate in horizontal directions, and it is not applicable for plane wave incidences.

3.1.3 Transmission line model for plane waves

To calculate the reflection phase of plane wave incidences at arbitrary incident angles and polarizations, a loaded transmission line model is established in [6–7], as shown in Fig. 3.4. The main idea is the decomposition of an EBG structure into a frequency selective surface (FSS) and a slab treated as a spacing medium between the FSS and the ground plane. When vertical vias are present in the mushroom-like EBG, the medium is treated as a so-called wire medium.

The EBG surface impedance Z_s results from two parts: the FSS grid impedance Z_g and the slab impedance Z_d. The value of Z_s is calculated from the parallel connection of Z_g and Z_d:

$$Z_s = \frac{Z_g Z_d}{Z_g + Z_d}. \tag{3.8}$$

The reflection coefficient is then computed as:

$$\Gamma^{TE} = \frac{Z_s \cos\theta - \eta_0}{Z_s \cos\theta + \eta_0}, \quad \Gamma^{TM} = \frac{Z_s - \eta_0 \cos\theta}{Z_s + \eta_0 \cos\theta}, \quad (3.9)$$

where η_0 is the free space wave impedance and θ is the incident angle. The resonant frequency for the zero degree reflection phase has to satisfy

$$X_g(\omega_0) + X_d(\omega_0) = 0, \quad (3.10)$$

where $X_g = \mathrm{Im}(Z_g)$, $X_d = \mathrm{Im}(Z_d)$.

As a metal backed wire medium, its impedance Z_d can be derived analytically. In most interesting cases ($k_d h \ll 1$), both TM and TE impedances are practically equal to [7]:

$$Z_d^{TE,TM} \approx j\omega\mu_0 h. \quad (3.11)$$

When vertical vias are absent, the TE impedance remains unchanged, but the TM impedance strongly depends on the incidence angle [7]:

$$Z_d^{TM} \approx j\omega\mu_0 h \cos^2\theta. \quad (3.12)$$

The grid impedance Z_g of the FSS depends on specific geometry used in the design. For an array of patches, its values are calculated as the following [7]:

$$Z_g^{TE}(\omega,\theta) = \frac{Z_g(\omega,0)}{\cos^2\theta}, \quad Z_g^{TM}(\omega,\theta) = Z_g(\omega,0)$$

$$Z_g(\omega,0) = \frac{1}{j\omega C_g}, \quad C_g = \frac{(1+\varepsilon_r)\varepsilon_0 a}{\pi} \log\left(\frac{2a}{\pi g}\right). \quad (3.13)$$

For other geometries such as Jerusalem cross FSS and spiral shape FSS, corresponding grid impedances are also derived using the series-resonant grid method [6]. The reflection phases obtained from this analytical model agree well with the numerical simulation data from Ansoft HFSS [7–8].

3.2 Graphic representation of frequency band gap

3.2.1 FDTD model

A unique feature of an electromagnetic band gap structure is the appearance of a frequency band where the propagation of electromagnetic waves is prohibited. In this section, the FDTD method is used to clearly visualize the frequency band gap feature. The computational code is based on a Cartesian grid cell with the absorbing boundary conditions [9]. An infinitesimal dipole source with a Gaussian pulse waveform is utilized to excite a wideband electromagnetic wave in the structure.

Figure 3.5 shows an FDTD simulation model: an infinitesimal dipole source surrounded by the mushroom-like EBG structure. The dipole is chosen to be vertically polarized because the E field is vertical to the ground plane in many applications such as microstrip antennas. As an example, two rows of EBG patches are plotted in Fig. 3.5. In FDTD simulations four rows, six rows, and eight rows of EBG patches are all simulated

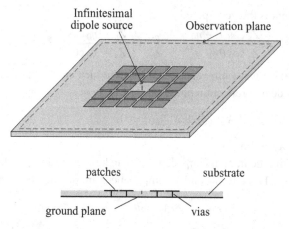

Fig. 3.5 An FDTD model to characterize frequency band gap of a finite EBG structure.

and compared. It is worthwhile to point out that the analyzed EBG structure here has a finite size, which is a practical case in real applications.

The EBG structure analyzed in this section has the following parameters:

$$W = 0.12\lambda_{6\,\text{GHz}}, \quad g = 0.02\lambda_{6\,\text{GHz}}, \quad h = 0.04\lambda_{6\,\text{GHz}}, \quad \varepsilon_r = 2.20, \quad (3.14)$$
$$r = 0.005\lambda_{6\,\text{GHz}}$$

where $\lambda_{6\,\text{GHz}}$ (50 mm) is the free space wavelength at 6 GHz. In the FDTD simulation, a uniform $0.02\,\lambda_{6\,\text{GHz}}$ (1 mm) discretization is used. The ground plane size is kept as $2.84\,\lambda_{6\,\text{GHz}} \times 2.84\,\lambda_{6\,\text{GHz}}$. An observation plane is positioned at $0.12\,\lambda_{6\,\text{GHz}}$ distance away from the edge, where it is located outside the EBG structure. The height of the reference plane is set to $0.04\,\lambda_{6\,\text{GHz}}$, which is equal to the substrate thickness. For the comparison purpose, a conventional case is also analyzed. This conventional (CONV.) case consists of a perfect electric conductor (PEC) ground plane and a dielectric substrate with the same thickness and permittivity as the EBG case.

The basic idea is to calculate and compare the E fields of two structures (EBG and CONV.) at the observation plane. Since the EBG structure can suppress the surface waves in a certain band gap, the E field outside the EBG structure should be weaker than that of the conventional case. To quantify the surface-wave suppression effect, an average $|\overline{E}|^2$ is calculated according to the following equation:

$$|\overline{E}|^2 = \frac{1}{S}\iint_S |E|^2 ds, \quad (3.15)$$

where S is the vertical observation plane whose boundary is plotted by the dashed line in Fig. 3.5.

Figure 3.6 plots the $|\overline{E}|^2$ of several EBG cases that are normalized to the $|\overline{E}|^2$ of the conventional case. The number of EBG rows is varied from two to eight to study the surface-wave suppression effect. It is observed that when fewer rows of EBG patches are used, the band gap effect is not significant. When the number of rows is increased, a clear band gap can be noticed. Inside this band gap, the average E field in the EBG case

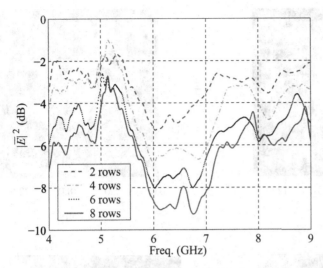

Fig. 3.6 Average E fields of the EBG case on the observation plane, which are normalized to the conventional case (from [10], © IEEE 2003).

is much lower than that in the conventional case. To determine the band gap region, a criterion is used that the average E field magnitude with the EBG is less than half of that without the EBG (the CONV. case), which is equivalent to the −6 dB level in Fig. 3.6. Thus a band gap from 5.8–7.0 GHz can be identified with a minimum of four rows of EBG patches.

3.2.2 Near field distributions inside and outside the frequency band gap

To visualize the band gap feature for surface-wave suppression, the near field distributions of the 8 row EBG case and the conventional case are calculated and graphically presented. Figure 3.7 plots the near field of both cases at 6 GHz, which is inside the band gap. The field level is normalized to 1 W delivered power and is shown in dB scale. The field level outside the EBG structure is around 10 dB. In contrast, the field level of the CONV. case is around 20 dB. The difference of field levels is due to the existence of the EBG structure, which suppresses the propagation of surface waves so that the field level in the EBG case is much weaker than in the conventional case.

However, the EBG structure cannot effectively suppress surface waves outside its frequency band gap. For example, Fig. 3.8 plots the near field of both cases at 5 GHz, which is outside the band gap according to Fig. 3.6. The field distribution of the CONV. case is similar to its distribution at 6 GHz. However, the field value outside the EBG structure is increased to around 20 dB, which is similar to that of the CONV. case. This means that although there are some interactions between the dipole source and the EBG structure, the field can still propagate through the EBG structure.

These near field distributions correlate well with the results in Fig. 3.6. From the comparison it can be concluded that there exists a frequency band gap region for the EBG structure where the propagation of surface waves is noticeably suppressed.

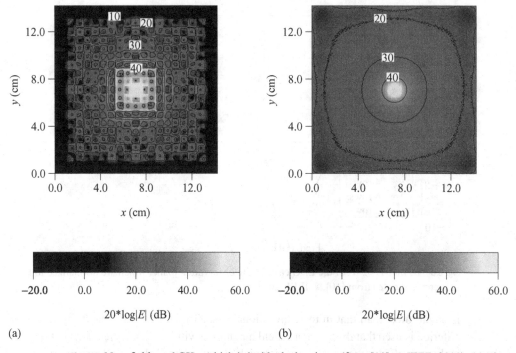

Fig. 3.7 Near fields at 6 GHz, which is inside the band gap (from [10], © IEEE, 2003). (a) The EBG case and (b) the CONV. case. The outside field of the EBG case is about 10 dB lower than that of the CONV. case because of the surface-wave suppression.

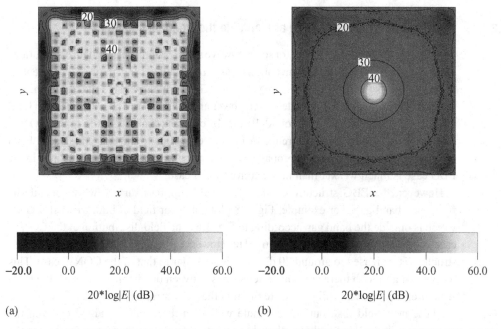

Fig. 3.8 Near fields at 5 GHz, which is outside the band gap (from [10], © IEEE 2003). (a) The EBG case and (b) the CONV. case. The outside field of the EBG case has a similar level as that of the CONV. case.

3.3 Frequency band gap for surface wave propagation

In the previous section, the FDTD method is used to analyze an EBG structure with a *finite size*. As the number of EBG cells increases, a clear surface wave band gap can be identified. However, it takes a long simulation time and an extensive storage memory to analyze a large structure using the previous FDTD model, which is not efficient in computation. Instead, with the utilization of periodic boundary conditions (PBCs), only a single cell of the EBG structure needs to be computed in the FDTD simulations. The PBCs are incorporated on four sides of the cell to model an infinite periodic replication, as shown in Fig. 2.6. This approach is computationally more efficient.

3.3.1 Dispersion diagram

Wavenumber k is an important parameter to describe the propagation property of electromagnetic waves. In a lossless case, the phase constant is $\beta = k$. Usually, β is a function of frequency ω. Once the phase constant is obtained, the phase velocity (u_p) and group velocity (u_g) can be derived [11]:

$$u_p = \frac{\omega}{\beta}, \quad u_g = \frac{d\omega}{d\beta}. \qquad (3.16)$$

Furthermore, the field distribution can also be determined, such as the field variation in a transverse direction. For a plane wave in free space, the relation between β and ω is a linear function:

$$\beta(\omega) = k = \omega\sqrt{\varepsilon_0\mu_0}. \qquad (3.17)$$

For surface waves propagating in a dielectric slab or an EBG structure, it is usually difficult to give an explicit expression for the wavenumber k. One has to either solve an eigen-value equation or perform a full wave simulation to determine the wavenumber. It is important to point out that the solution of an eigen-value equation may not be unique. In another words, there may exist several different propagation constants at the same frequency. Each one is known as a specific mode with its own phase velocity, group velocity, and field distribution. The relation between β and ω is often plotted out and referred to as the dispersion diagram.

For a periodic structure such as the EBG, the field distribution of a surface wave is also periodic with a proper phase delay determined by the wavenumber k and periodicity p [12]. Thus, each surface wave mode can be decomposed into an infinite series of space harmonic waves, as discussed in Section 2.3.2,

$$\vec{E}(x,y,z) = \sum_{n=-\infty}^{\infty} \vec{E}_n(y,z)e^{-jk_{xn}x}, \quad k_{xn}(\omega) = k_x(\omega) + n\frac{2\pi}{p}. \qquad (3.18)$$

Here, we assume the periodic and propagation direction is the x direction. Although these space harmonics have different phase velocities, they share the same group velocity. Furthermore, these space harmonics cannot exist individually because each single harmonic does not satisfy the boundary conditions of the periodic structure. Only their summation satisfies the boundary conditions. Thus, they are considered to be the same mode.

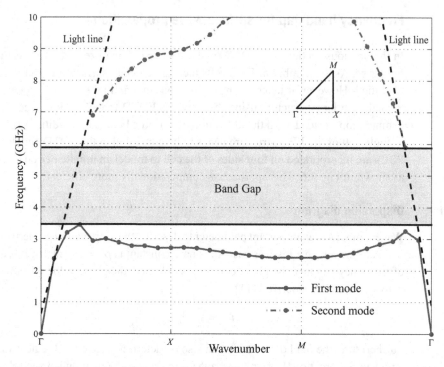

Fig. 3.9 FDTD simulated dispersion diagram of the mushroom-like EBG structure. A clear band gap for surface waves can be observed.

Another important observation from (3.18) is that the dispersion curve $k_x(\omega)$ is periodic along the k axis with a periodicity of $2\pi/p$ [13]. Therefore, we only need to plot the dispersion relation within one single period, namely, $0 \leq k_{xn} \leq 2\pi/p$, which is known as the Brillouin zone. This concept can be easily extended to two-dimensional periodic structures, where the Brillouin zone becomes a two-dimensional square area:

$$0 \leq k_{xn} \leq 2\pi/p_x, \quad 0 \leq k_{yn} \leq 2\pi/p_y. \tag{3.19}$$

3.3.2 Surface wave band gap

Based on the concepts discussed above, we start to characterize the dispersion diagram of the mushroom-like EBG structure using the FDTD/PBC method. The dimensions of the analyzed EBG structure are:

$$W = 0.10\lambda, \quad g = 0.02\lambda, \quad h = 0.04\lambda, \quad \varepsilon_r = 2.94, \quad r = 0.005\lambda. \tag{3.20}$$

The free space wavelength at 4 GHz, $\lambda = 75$ mm, is used as a reference length to define the physical dimensions of the EBG structure. These parameters are selected for antenna applications that will be discussed in the following chapters, and they are readily scaled to other frequencies of interest.

Figure 3.9 shows the dispersion diagram of the mushroom-like EBG structure. The vertical axis shows the frequency and the horizontal axis represents the values of the transverse wavenumbers (k_x, k_y). Three specific points are:

$$\Gamma: \quad k_x = 0, \quad k_y = 0, \qquad (3.21)$$

$$X: \quad k_x = \frac{2\pi}{(W+g)}, \quad k_y = 0, \qquad (3.22)$$

$$M: \quad k_x = \frac{2\pi}{(W+g)}, \quad k_y = \frac{2\pi}{(W+g)}, \qquad (3.23)$$

From point Γ to point X, k_x increases while k_y is zero. From point X to point M, k_x is fixed to $2\pi/(W+g)$ while k_y increases from zero to $2\pi/(W+g)$. From point M to point Γ, both k_x and k_y decrease from $2\pi/(W+g)$ to zero. In summary, the horizontal axis varies along a triangle boundary of the Brillouin zone shown in the figure. Note that the other triangle parts in the square Brillouin zone will have the same propagation constant behavior because of the symmetry of the geometry.

In the FDTD simulation, we are trying to identify the eigen-frequencies for specific wavenumbers (k_x, k_y). The FDTD simulation is repeated for thirty different combinations of k_x and k_y in the horizontal axis, and the resonant frequencies of surface waves are extracted and plotted in Fig. 3.9. Each point in the modal diagram represents a certain surface wave mode. Connecting the modes with similar field distributions together, we obtain dispersion curves of the EBG structures.

The first mode starts at zero frequency and the eigen-frequency increases with the wavenumber. When it reaches a turning point at 3.5 GHz, it decreases as the wavenumber is further increased. In the FDTD simulation, it is also noticed that the field for the first mode is TM dominant. The second mode starts to appear at frequency higher than 5.9 GHz. It is clear that no eigen-mode exists in the frequency range from 3.5 GHz to 5.9 GHz. Thus, this frequency region is defined as a surface wave band gap. No surface waves can propagate in the EBG structure inside this frequency band gap.

3.4 In-phase reflection for plane wave incidence

3.4.1 Reflection phase

Besides the surface wave property, EBG structures also exhibit interesting reflection phase behavior, which will be discussed in this section. Reflection coefficient is a popular parameter used to describe the reflection property of an object. It is defined as the ratio of the reflected field over the incident field at the reflecting surface. Usually, it has a complex value with the corresponding magnitude and phase. When a ground plane exists in an analyzed lossless structure, the magnitude is always one because all the energy is reflected back. In this case, the *reflection phase* is of special interest. In practice, the reflection phase has been used in the designs of many antennas and microwave devices, such as reflectarray elements and polarizers.

As a starting point, let's start with the traditional conductors. If a plane wave is normally impinged upon a perfect electric conductor (PEC), the total tangential E field must be zero in order to satisfy the boundary condition. Thus, the reflected E field and the incident

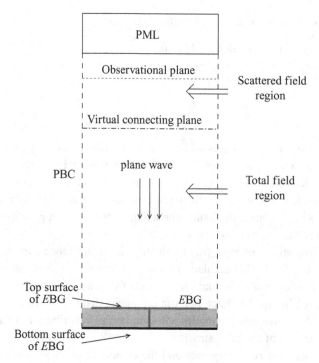

Fig. 3.10 An FDTD model of the EBG structure for reflection phase characterization (from [14], © IEEE 2003).

E field should have the opposite signs, resulting in a reflection coefficient of -1. The reflection phase is 180° for the PEC case. For a perfect magnetic conductor (PMC), the total tangential H field must be zero. Thus, the reflected H field and the incident H field should have the opposite signs whereas the reflected E field and the incident E field have the same signs. As a result, the reflection coefficient is equal to one and the corresponding reflection phase is 0° for PMC case. However, the PMC surface does not exist in nature.

Research on EBG structures reveals that they can realize the PMC condition in a certain frequency band. Thus, they are sometimes referred to as artificial magnetic conductors (AMC). Actually, the reflection phase of an EBG structure is a function of frequency. It varies continuously from 180° to $-180°$ as frequency increases. This reflection phase feature makes EBG surfaces unique and initiates many low profile wire antenna designs on EBG surfaces.

3.4.2 EBG reflection phase: normal incidence

To evaluate the reflection phase of EBG structures, the FDTD method discussed in Chapter 2 is used here in the analysis. Figure 3.10 shows a unit cell model for the computation of the reflection phase at normal incidence [14]. A single unit of the EBG structure with periodic boundary conditions (PBC) on four sides is simulated to model an infinite periodic structure. The perfectly matched layers (PML) are positioned 0.55 λ above the EBG to absorb the reflected energy. The total field/scattered field formulation is used to incorporate the plane wave excitation into the computational domain. The normally incident plane wave is launched on a virtual connecting surface that is located 0.40 λ above the EBG bottom surface. An observation plane is located on the scattered field region to record the reflected field from the EBG surface, and the height of the observation plane is 0.50 λ.

No matter in FDTD simulations or in the practical measurements, the observation plane is usually set away from the EBG surface. The reason is the existence of high-order harmonics near the EBG surface. When the observation plane is selected near the EBG surface, the results are always affected by the higher-order modes. These higher-order harmonics have a large wavenumber along the periodic directions, and the corresponding wavenumber in the vertical direction is usually imaginary. As a result, the field strengths of the higher-order modes decay exponentially along the vertical direction. When the observational plane is set away from the EBG surface, their effects can be minimized.

Because of the different locations of the observation plane and reflecting EBG surface, a question is how to restore the reflection phase exactly on the EBG surface. To solve this problem, an ideal PEC surface is used as a reference. The scattered fields from a PEC surface are also calculated using the FDTD method. Note that the PEC surface is located at the same height as the EBG top surface while the observational plane stays the same. Then, the reflected phase from the EBG structure is normalized to the reflected phase from the PEC surface:

$$\phi = \phi^{EBG} - \phi^{PEC} + \pi \qquad (3.24)$$

Therefore, the propagation phase from the distance between the reflecting surface and observation plane is canceled out. A factor of π is added to the phase result to account for the reference of the PEC surface, which is known to have a reflection phase of π radians. The EBG reflection phase characterization obtained from the above procedure follows the same methodology applied in the measurements [1].

The same EBG structure as (3.20) is also analyzed here for plane wave reflection. Figure 3.11 shows the FDTD computed reflection phase curve. It is observed that the reflection phase of the EBG surface decreases continuously from 180° to −180° as frequency increases. At low-frequency and high-frequency regions, the EBG surface shows a similar phase to a PEC case, which is 180° (or −180°). At frequencies around 5.74 GHz, the EBG surface exhibits a reflection phase close to 0°, which resembles a PMC surface. In addition, other reflection phases can be also realized by the EBG surface. For example, a 90° reflection phase is achieved around 4.82 GHz.

3.4.3 EBG reflection phase: oblique incidence

The reflection phase of an EBG structure varies with the incident angle and polarization state. In this section, we present the reflection phase results for oblique incidences at both TE and TM polarizations. Both the split-field FDTD method [15] and the constant-k_x FDTD method [16] are used here to calculate the oblique incidence problem.

For the TE incident case, the electric field is set along the y direction and the plane wave is incident on the xz plane. The incident angle, defined as the angle between the propagation vector and the z axis, varies from 0 to 90 degrees. The EBG dimensions

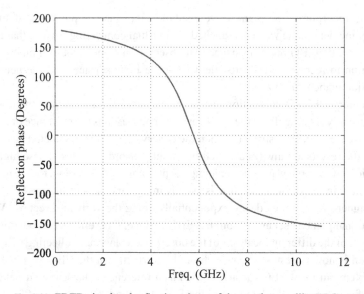

Fig. 3.11 FDTD simulated reflection phase of the mushroom-like EBG structure for normal incidence. A zero degree reflection phase is obtained at 5.74 GHz.

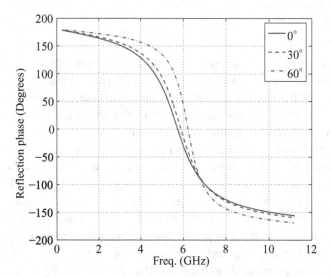

Fig. 3.12 FDTD simulated reflection phase of the mushroom-like EBG structure for the TE incidences at different incident angles.

are the same as before and Fig. 3.12 compares the results for 0, 30, and 60 degrees of incidence. When the incident angle increases, the resonant frequency (where the reflection phase is equal to zero) also slightly increases from 5.74 GHz for normal incidence to 5.89 GHz for 30° incidence and 6.23 GHz for 60° incidence. It is also noticed that the slope near the resonance becomes steep when the incidence angle increases.

3.4 In-phase reflection for plane wave incidence

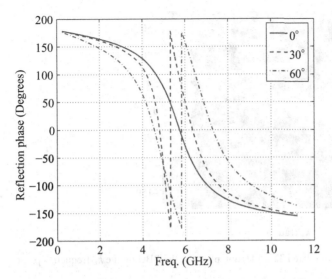

Fig. 3.13 FDTD simulated reflection phase of the mushroom-like EBG structure for the TM incidences at different incident angles.

For the TM incident case, the magnetic field is set along the y direction and the plane wave is also incident on the xz plane. Figure 3.13 shows the FDTD simulated results for 0, 30, and 60 degrees of incidence. An interesting dual-resonance behavior is observed on the reflection phase curves. As the incident angle of the TM wave is offset from the normal direction, one resonance appears at a frequency lower than the original frequency and the other higher. When the incident angle increases, the lower resonant frequency decreases from 4.80 GHz (30°) to 4.56 GHz (60°) whereas the higher resonant frequency increases from 6.34 GHz (30°) to 7.28 GHz (60°). The frequency separation between the two resonances increases as the incident angle increases.

An important reason for the different TE and TM results is the existence of vertical vias located in the center of EBG patches. For the TE incidence, the electric field is perpendicular to the vias. Thus, the boundary condition does not change, and the center vias have no effect for the TE incidence cases. The reflection phase is only determined by the induced current on the metal patch.

For the obliquely incident TM wave, the E field has a vertical component parallel to the vias. Thus, the boundary condition needs to be enforced on the vias, and an electric current is induced on the vertical vias. The current magnitude changes with the incidence angle. Now the reflection phase is determined both by the induced current on the patch and by the induced current on the vias. Their coupling effect forms two resonances in Fig. 3.13.

The reflection phase behaviors of oblique incidence can also be clearly visualized in the k_x-frequency plane. Using the constant k_x approach for periodic boundary conditions, the FDTD simulation is repeated for 101 different k_x values uniformly sampled from 0 to 251.3 radian/m. The reflection phases are calculated at each frequency and k_x. The

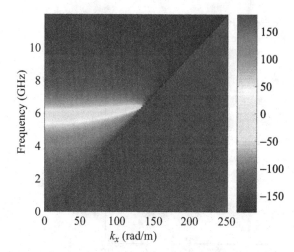

Fig. 3.14 FDTD simulated TE reflection phase of the EBG on the k_x-frequency plane.

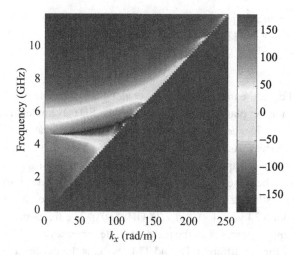

Fig. 3.15 FDTD simulated TM reflection phase of the EBG on the k_x-frequency plane.

results for TE incidence are plotted in Fig. 3.14 and TM incidence in Fig. 3.15. Note we are only interested in the plane wave region, where

$$k_0 = 2\pi f/C \geq k_x. \tag{3.25}$$

For the TE incidence, as the k_x increases, the resonant frequency slightly increases. In addition, the region around 0 degrees becomes narrower. For the TM incidence, a bifurcation phenomenon is clearly observed. This corresponds to the dual resonance behavior in Fig. 3.13.

3.5 Soft and hard surfaces

Electromagnetic band gap structures also demonstrate properties referred to as soft and hard operations [17–18]. In general, a periodic ground plane may be built to either stop the propagation of waves or support a quasi-transverse electromagnetic wave along the surface. The former is called a soft surface, and the latter is a hard surface. These definitions are analogs to the acoustic terminologies. There are many applications of artificially soft and hard surfaces. For example, a horn antenna with hard walls supports transverse electromagnetic (TEM) waves, which results in a uniform aperture distribution and zero taper [19].

Usually, a soft or a hard surface can only be realized in certain frequency and wavenumber regions. Therefore, it is critical to accurately identify these regions so that the surface can be used effectively. In this section, a clear definition for soft and hard operations is provided using the impedance and reflection coefficient of the surface. An associated bandwidth definition is also suggested. An advantage of these definitions is to *provide a unified approach to deal with both surface wave and plane wave incidences* [20]. To demonstrate the concept, the reflection coefficients and impedances of a corrugated surface and a mushroom-like EBG are calculated, and the bandwidths for soft and hard behaviors are specified.

3.5.1 Impedance and reflection coefficient of a periodic ground plane

An electromagnetic wave incident on a periodic ground plane with specific wavenumbers (k_x, k_y, k_z), frequency (k_0), and the time convention $e^{j\omega t}$ can be written as follows:

$$\begin{cases} \vec{E}^i = (E^i_{TM}\hat{l} + E^i_{TE}\hat{t} + E^i_z\hat{z})e^{jk_xx+jk_yy}e^{jk_zz} \\ \vec{H}^i = (H^i_{TE}\hat{l} + H^i_{TM}\hat{t} + H^i_z\hat{z})e^{jk_xx+jk_yy}e^{jk_zz} \end{cases} \quad (3.26)$$

For the TM_z wave, the electric field has longitudinal and z components while the magnetic field is along the transverse direction. For the TE_z wave, the magnetic field has longitudinal and z components while the electric field is along the transverse direction. The periodicity is a along the x direction and b along the y direction.

The total tangential field in the free space above the periodic ground plane can be represented as the sum of incident and reflected waves:

$$\begin{cases} \vec{E}_{\tan} = \vec{E}^i_{\tan} + \vec{E}^r_{\tan} = \vec{E}^{0,0}_{\tan} + \sum_m \sum_n (E^{m,n}_{TM}\hat{l} + E^{m,n}_{TE}\hat{t}) \times e^{j(m\frac{2\pi}{a}x+n\frac{2\pi}{b}y)}e^{-jk^{m,n}_z z} \\ \vec{H}_{\tan} = \vec{H}^i_{\tan} + \vec{H}^r_{\tan} = \vec{H}^{0,0}_{\tan} + \sum_m \sum_n (H^{m,n}_{TE}\hat{l} + H^{m,n}_{TM}\hat{t}) \times e^{j(n\frac{2\pi}{a}x+m\frac{2\pi}{b}y)}e^{-jk^{m,n}_z z} \end{cases}$$

$$\left(k_x + m\frac{2\pi}{a}\right)^2 + \left(k_y + n\frac{2\pi}{b}\right)^2 + (k^{m,n}_z)^2 = k_0^2, \quad m \cdot n \neq 0. \quad (3.27)$$

The tangential components are expanded into a series of Floquet space harmonics. For the zero harmonic, it includes both the incident and reflected components:

$$\begin{cases} \vec{E}^{0,0}_{\tan} = (E^i_{TM}e^{jk_zz} + E^{0,0}_{TM}e^{-jk_zz})\hat{l} + (E^i_{TE}e^{jk_zz} + E^{0,0}_{TE}e^{-jk_zz})\hat{t} \\ \vec{H}^{0,0}_{\tan} = (H^i_{TE}e^{jk_zz} + H^{0,0}_{TE}e^{-jk_zz})\hat{l} + (H^i_{TM}e^{jk_zz} + H^{0,0}_{TM}e^{-jk_zz})\hat{t} \end{cases} \quad (3.28)$$

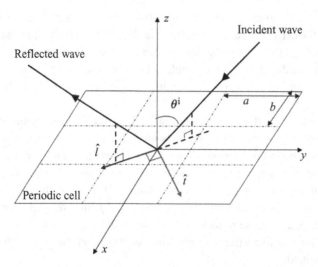

Fig. 3.16 An electromagnetic wave incident on a periodic ground plane (from [20], © IEEE 2005). The incident angle θ^i, longitudinal direction $\hat{l}$, and transverse direction $\hat{t}$ are labeled in the figure.

With field components defined above, we now formulate the impedance and the reflection coefficient of a periodic ground plane. The TM (or TE) impedance of a surface is defined as the ratio of the tangential electric field over the tangential magnetic field right on the surface:

$$E^i_{TE} = H^i_{TE} = 0 \rightarrow Z_{TM} = \frac{E^i_{TM} + E^{0,0}_{TM}}{H^i_{TM} + H^{0,0}_{TM}} \tag{3.29}$$

$$E^i_{TM} = H^i_{TM} = 0 \rightarrow Z_{TE} = \frac{E^i_{TE} + E^{0,0}_{TE}}{H^i_{TE} + H^{0,0}_{TE}} \tag{3.30}$$

Note that the zero harmonic is considered here. The orthogonal polarization component, TE (or TM) part of the incident field, is set to zero in the calculation.

The TM and TE reflection coefficients for the surface are defined as follows using the zero harmonic:

$$E^i_{TE} = H^i_{TE} = 0 \rightarrow R_{TM} = \frac{H^{0,0}_{TM}}{H^i_{TM}} = -\frac{E^{0,0}_{TM}}{E^i_{TM}} \tag{3.31}$$

$$E^i_{TM} = H^i_{TM} = 0 \rightarrow R_{TE} = \frac{E^{0,0}_{TE}}{E^i_{TE}} = -\frac{H^{0,0}_{TE}}{H^i_{TE}} \tag{3.32}$$

The reflection coefficient can also be derived from the surface impedance and the wave impedances:

$$R_{TM} = -\frac{Z_{TM} - Z^{wave}_{TM}}{Z_{TM} + Z^{wave}_{TM}} \tag{3.33}$$

$$R_{TE} = \frac{Z_{TE} - Z^{wave}_{TE}}{Z_{TE} + Z^{wave}_{TE}}. \tag{3.34}$$

3.5 Soft and hard surfaces

Table 3.1 Wave parameters in the plane wave and surface wave regions

	Identification	Wave impedance	Surface impedance	Reflection coefficient
Plane wave region	$k_x^2 + k_y^2 < k_0^2$, k_z Real	Real	Imaginary	Complex, $\|R\| = 1$
Surface wave region	$k_x^2 + k_y^2 > k_0^2$, k_z Imaginary	Imaginary	Imaginary	Real

where Z_{TM}^{wave} and Z_{TE}^{wave} are the wave impedances for TM and TE waves, and can be written as follows [20]:

$$Z_{TM}^{wave} = \frac{E_{TM}^i}{H_{TM}^i} = -\frac{E_{TM}^{0,0}}{H_{TM}^{0,0}} = \frac{k_z}{k_0}\eta_0 \qquad (3.35)$$

$$Z_{TE}^{wave} = \frac{E_{TE}^i}{H_{TE}^i} = -\frac{E_{TE}^{0,0}}{H_{TE}^{0,0}} = \frac{k_0}{k_z}\eta_0. \qquad (3.36)$$

Table 3.1 compares wave impedances, surface impedances, and reflection coefficients in the plane wave and surface wave region. As can be seen, for a plane wave incidence $(k_x^2 + k_y^2 < k_0^2)$, k_z is real; therefore, the wave impedance is real based on (3.35) and (3.36). For a lossless ground plane, the surface impedance is purely imaginary and has no real part. Thus, one can derive from (3.33) and (3.34) that the magnitude of the reflection coefficient is always unity in this region and its phase varies with the wavenumber and frequency of the incident wave as shown in Fig. 3.17.

For an incidence wave in the surface wave region $(k_x^2 + k_y^2 > k_0^2)$, k_z is imaginary so the wave impedance is also imaginary. This results in a real value for the reflection coefficient as shown in Fig. 3.17. In contrast to the plane wave incidence, the magnitude of the reflection coefficient in this case changes with the wavenumber and frequency of the incident wave whereas the phase is fixed at 0° or 180°.

3.5.2 Soft and hard operations

From the discussion above, the reflection coefficient is a function of both wavenumbers (k_x, k_y) and frequency (k_0) of the incident wave. When a surface is used for soft or hard operation, its reflection coefficient must satisfy a specific condition, which is derived here. Furthermore, a bandwidth is defined for the soft and hard operations. The definition gives a unified approach for the identification of the bandwidth in both the plane wave and the surface wave regions.

Bandwidth for soft and hard operations

A soft operation for TM (or TE) waves means the total magnetic (or electric) field is zero in the transverse $\hat{t}$ direction [17]. This condition results in zero power density flux along the surface, i.e., the Poynting vector is zero in the direction $\hat{l}$. From the reflection coefficient definitions in (3.31) and (3.32), an ideal soft operation is described as follows:

$$R = -1 \qquad (3.37)$$

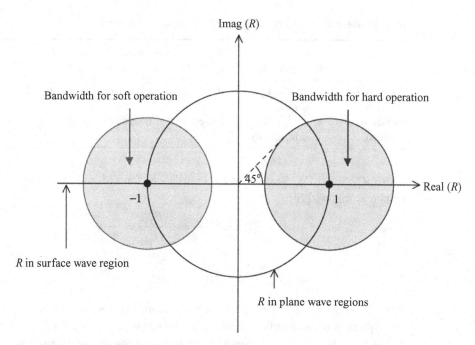

Fig. 3.17 Complex R plane and the bandwidth circles for soft and hard operations (from [20], © IEEE 2005). In the plane wave region (visible region), R varies on the unit circle and in the surface wave region R varies along the real axis.

where -1 reflection gives the zero field on the surface in the direction $\hat{t}$ as needed for the soft operation. Next, we define a bandwidth for the soft operation as a wavenumber-frequency region where the reflection coefficient satisfies the following requirement:

$$\mathrm{BW_s} = \left\{ (k_x, k_y, k_0) \quad | \quad F^\mathrm{s} = |R(k_x, k_y, k_0) - (-1)| < \frac{1}{\sqrt{2}} \right\} \qquad (3.38)$$

in which F^s is a soft factor which has to be less than $1/\sqrt{2}$ ($F^\mathrm{s}_\mathrm{dB} < -3$ dB) for the soft operation.

To illustrate the bandwidth concept above, Fig. 3.17 shows the reflection coefficient in the complex plane. According to Table 3.1, R of a lossless ground plane has a complex value with the magnitude of 1 in the plane wave region. In the surface wave region, R is real but its magnitude varies. The soft bandwidth definition restricts the complex value of R in a circle with center at -1 and radius of $1/\sqrt{2}$. The selection of $1/\sqrt{2}$ as radius guarantees that the phase of reflection coefficient differs less than $45°$ from the ideal $180°$ of -1 reflection in the plane wave region. It also restricts the maximum and minimum values of the reflection in the surface wave region between $-1 - 1/\sqrt{2}$ and $-1 + 1/\sqrt{2}$. In summary, it provides a unified approach for the characterization of bandwidth in both the plane wave and the surface wave regions.

Hard operation for TM (or TE) waves means zero magnetic (or electric) fields in the longitudinal direction $\hat{l}$ [17]. This condition can be obtained by setting the reflection

3.5 Soft and hard surfaces

Table 3.2 Reflection coefficient and surface impedance for soft, hard, PEC, and PMC surfaces.

	TM	TE
Soft surface	$R_{TM} = -1$, $Z_{TM} = \infty$ (Soft)	$R_{TE} = -1$, $Z_{TE} = 0$ (Soft)
Hard surface	$R_{TM} = 1$, $Z_{TM} = 0$ (Hard)	$R_{TE} = 1$, $Z_{TE} = \infty$ (Hard)
PEC	$R_{TM} = 1$, $Z_{TM} = 0$ (Hard)	$R_{TE} = -1$, $Z_{TE} = 0$ (Soft)
PMC	$R_{TM} = -1$, $Z_{TM} = \infty$ (Soft)	$R_{TE} = 1$, $Z_{TE} = \infty$ (Hard)

coefficient to 1 as follows:

$$R = 1. \tag{3.39}$$

As shown in Fig. 3.17, we define the bandwidth for hard operation as a wavenumber-frequency region where the reflection coefficient satisfies the following requirement:

$$BW_h = \left\{ (k_x, k_y, k_0) \mid F^h = |R(k_x, k_y, k_0) - 1| < \frac{1}{\sqrt{2}} \right\}. \tag{3.40}$$

F^h is the hard factor which has to be less than $1/\sqrt{2}$ ($F^h_{dB} < -3$ dB) for the hard operation. This condition restricts the value of R in a circle with center at 1 and radius of $1/\sqrt{2}$. This definition is also applicable in both the plane wave and the surface wave regions.

Surface wave modes

Surface wave mode is a natural resonance in the structure that can exist without excitation. Thus, a surface wave mode can be interpreted as the non-zero scattered field for the zero incident wave. This condition is equivalent to an infinite value for the reflection coefficient R. To obtain an infinite value for the reflection coefficient in (3.33) and (3.34), surface impedance must be opposite to the wave impedance. Thus, the dispersion curve for surface wave modes in the wavenumber-frequency plane can be identified as follows:

$$\text{Dispersion curve} = \left\{ (k_x, k_y, k_0) \mid R(k_x, k_y, k_0) = \infty, \ Z^{surf} = -Z^{wave} \right\} \tag{3.41}$$

A TM (or TE) mode occurs when the TM (or TE) reflection coefficient grows to infinity. A hybrid mode happens when the surface has infinite reflection coefficient for both the TM and TE waves.

Soft, hard, PEC, and PMC surfaces

Ideal soft and hard operations are defined as having the reflection coefficients of -1 and 1 respectively. According to the relation between the reflection coefficient and surface impedance in (3.33) and (3.34), one can conclude that a high TM surface impedance and a low TE surface impedance result in the reflection of -1, namely, the soft operation of the surface. In contrast, low TM impedance and high TE impedance give the reflection coefficient of 1 and the hard operation for TM and TE waves.

As summarized in Table 3.2, a soft surface has soft operation for both TM and TE waves. Similarly, a hard surface has hard operation for both polarizations. A perfect electric conductor (PEC) has zero TM and TE surface impedances; therefore it operates

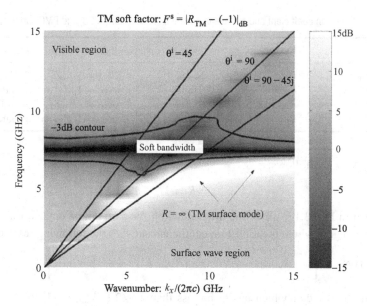

Fig. 3.18 TM soft factor of a corrugated surface shown in Fig. 2.24. Wave propagates along the x direction (from [20], © IEEE 2005).

hard for TM waves and soft for TE waves. A perfect magnetic conductor (PMC) has infinite TM and TE impedances, and it operates soft for TM waves and hard for TE waves.

3.5.3 Examples

Corrugated surface

To demonstrate the concepts discussed above, two examples are provided here. First one is a corrugated surface, where periodic metal bars are added to a dielectric slab. The same structure is analyzed at the end of Chapter 2 using the FDTD/PBC method and the TM and TE impedances are obtained in Figs. 2.25 and 2.26.

When the wave propagates along the x direction, low TE impedance and high TM impedance are obtained, resulting in a soft operation. The reflection coefficients can be computed from (3.33) and (3.34) and the soft factor is calculated from (3.38). Fig. 3.18 shows the TM soft factor and the corresponding -3dB contour. Inside this contour, the soft factor falls below -3dB, resulting in a soft operation for TM waves. For the TE case, the low surface impedance leads to a reflection coefficient very close to -1. Thus, the soft factor for TE waves is below -3dB over the entire frequency-wavenumber plane. In summary, the bandwidth for the surface to operate soft for both TM and TE waves is the region inside the -3dB contour in Fig. 3.18.

In the wavenumber-frequency plane, three lines corresponding to the plane wave incident with $45°$, $90°$ (grazing incidence, light line), $90°-45°\phi$ (surface wave incidence) are also labeled in Fig. 3.18. It is noticed that a soft surface is realized in a certain frequency band for all incident angles. It is also interesting to observe that the bandwidth

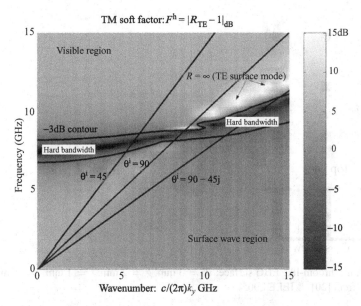

Fig. 3.19 TE hard factor of a corrugated surface shown in Fig. 2.24. Wave propagates along the y direction (from [20], © IEEE 2005).

for soft operation broadens along the light line, showing that a soft surface operates efficiently for the grazing incidence.

The region with infinite reflection is identified as the region for propagation of a surface wave mode. This region is located close to the 15dB contour in Fig. 3.18, which shows a high reflection coefficient. As can be seen, the soft operation for surface occurs for the frequencies above the propagation of the TM surface wave mode.

When the wave propagates along the y direction, high TE impedance and low TM impedance are observed, which indicates a hard operation. The hard factor for TE waves is calculated accordingly and plotted in Fig. 3.19. The -3dB contour is also specified in this figure to indicate hard operation. An area with infinite reflection is observed at this figure which represents a TE surface wave mode.

The region for hard operation occurs at the frequencies below the TE surface wave. The -3dB contour shifts toward the higher frequencies when the wavenumber of the incident wave is increased. This indicates that as one increases the angle of incidence from the normal direction, the frequency band for the hard operation shifts toward higher frequencies. It can be seen from this figure that there is no specific frequency band that the surface operates as a hard surface for all wavenumbers. It is also interesting to note that the region for hard operation narrows down along the light line, indicating that the surface does not operate efficiently near the grazing incidence.

For the TM waves, the low surface impedance leads to a reflection coefficient close to 1. Thus, the TM hard factor is below -3dB at entire wavenumber-frequency plane. Therefore, the bandwidth of the surface to operate hard for both TM and TE waves is the region inside the -3dB contour of the TE hard factor plot.

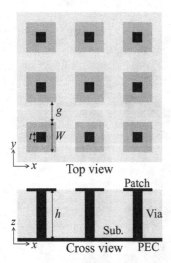

Fig. 3.20 A mushroom-like EBG surface: $W = 3$ mm, $g = 2$ mm, $t = 1$ mm, $h = 5$ mm, $\varepsilon_r = 4.0$ (from [20], © IEEE 2005).

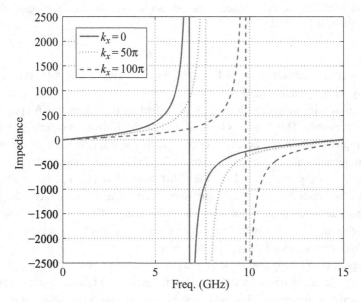

Fig. 3.21 TE surface impedance of the EBG structure.

Mushroom-EBG surface

It has been revealed that a mushroom-like EBG can be used to realize an artificial PMC. As shown in Table 3.2, a PMC surface has soft operation for the TM wave and hard operation for the TE wave. To illustrate these features, a mushroom-like EBG shown in Fig. 3.20 is analyzed. The dimensions are provided in the caption. The TM and TE impedances of the EBG are calculated using the unified FDTD/PBC method discussed in Section 2.5 and the results are plotted in Figs. 3.21 and 3.22. Both TM and TE impedances

3.5 Soft and hard surfaces

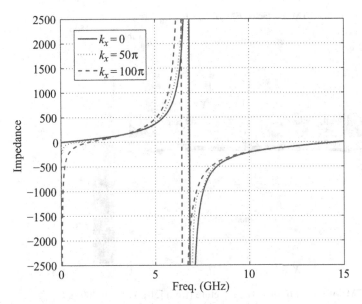

Fig. 3.22 TM surface impedance of the EBG structure.

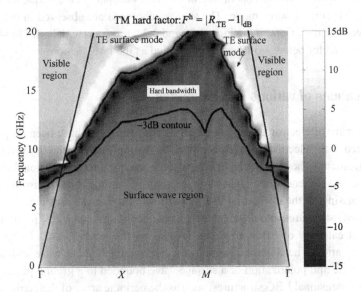

Fig. 3.23 TE hard factor of the EBG structure (from [20], © IEEE 2005).

show high values at a certain frequency. The TM impedance is almost independent of wavenumber, whereas the TE impedance varies with the wavenumber.

The hard factor for the TE waves and soft factor for TM are plotted in Figs. 3.23 and 3.24, respectively. The region with infinite reflection represents the TM and TE surface wave modes. The -3dB contours are plotted in these figures. It is observed that the soft operation for TM waves happens at the frequencies slightly above the TM surface

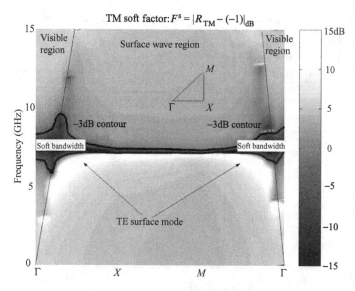

Fig. 3.24 TM soft factor of the EBG structure (from [20], © IEEE 2005).

wave mode, whereas the hard operation for TE waves happens at the frequencies slightly below the TE surface wave mode. The same phenomena are observed in the previous corrugated structure. Combining Figs. 3.23 and 3.24, the areas with both TM soft factor and TE hard factor below −3dB can be identified as the PMC region.

3.6 Classifications of various EBG structures

A wide variety of electromagnetic band gap (EBG) structures have been proposed and investigated in the electromagnetics community. A simple way to classify them is based on their geometry. They can be classified into three groups: three-dimensional volumetric structures, two-dimensional planar surfaces, and one-dimensional transmission lines. After discussing all the interesting EM properties in the previous sections, now we can classify EBG structures into different categories according to their EM properties [21].

The first category of EBG structures focuses on the inhibition of electromagnetic wave propagation. The electromagnetic wave can be either a plane wave with a specific incident angle and polarization or a surface wave bounded to a ground plane. Most of the three-dimensional EBG structures, such as the periodic array of dielectric rods [22], fall into this category. Some two-dimensional surfaces can also be put into this category when the surface waves are prohibited.

The second category of EBG structures emphasizes the reflection phase property. Usually two-dimensional surfaces with a very thin profile are being considered in this category. For example, some frequency selective surfaces backed by a thin grounded slab are investigated and their reflection phases are characterized [23–24].

It is worthwhile to point out that these two categories have many overlaps. For example, the mushroom-like EBG [1] and the uni-planar EBG [5] belong to both groups. However, there also exist some differences between the two groups. Some in-phase reflecting surfaces may not have a band gap for surface waves. An example is a periodic patch loaded slab, which will be discussed in detail in Chapter 7.

As a final remark in this chapter, we would like to highlight the importance of wave properties, namely, the propagation direction and polarization. Although there are several terminologies in the literature, such as EBG, artificial magnetic conductor (AMC), high impedance surfaces (HIS), they are always used to describe the structure's electromagnetic behavior under specific wave direction and polarization. It is difficult, or maybe not necessary, to give a unified definition for all of them. However, readers are reminded to be cautious when applying them in specific applications.

3.7 Project

Project 3.1 For a grounded dielectric slab, derive the following properties with different polarizations and propagation directions:

(a) dispersion curves
(b) reflection phases
(c) soft and hard factors.

References

1. D. Sievenpiper, L. Zhang, R. F. J. Broas, N. G. Alexopolus, and E. Yablonovitch, "High-impedance electromagnetic surfaces with a forbidden frequency band," *IEEE Trans. Microwave Theory Tech.*, **vol. 47**, 2059–74, 1999.
2. D. F. Sievenpiper, *High-Impedance Electromagnetic Surfaces*, Ph.D. dissertation at University of California, Los Angeles, 1999.
3. M. Rahman and M. A. Stuchly, "Transmission line – periodic circuit representation of planar microwave photonic bandgap structures," *Microwave and Optical Tech. Lett.*, **vol. 30, no. 1**, 15–19, 2001.
4. D. M. Pozar, *Microwave Engineering*, Wiley, 1998.
5. R. C. Coccioli, F. R. Yang, K. P. Ma, and T. Itoh, "Aperture-coupled patch antenna on UC-PBG substrate," *IEEE Trans. Microwave Theory Tech.*, **vol. 47**, 2123–30, 1999.
6. S. Maci, M. Caiazzo, A. Cucini, and M. Casaletti, "A pole-zero matching method for EBG surfaces composed of a dipole FSS printed on a grounded dielectric slab," *IEEE Trans. Antennas Propagat.*, **vol. 53, no. 1**, 70–81, 2005.
7. C. R. Simovski, P. Maagt, and I. V. Melchakova, "High impedance surfaces having resonance with respect to polarization and incident angle," *IEEE Trans. Antennas Propagat.*, **vol. 53, no. 3**, 908–14, 2005.
8. G. Gampala, *Analysis and design of artificial magnetic conductors for X-band antenna applications*, M.Sc. thesis at The University of Mississippi, 2007.
9. M. A. Jensen, *Time-Domain Finite-Difference Methods in Electromagnetics: Application to Personal Communication*, Ph.D. dissertation at University of California, Los Angeles, 1994.

10. F. Yang and Y. Rahmat-Samii, "Microstrip antennas integrated with electromagnetic band-gap (EBG) structures: a low mutual coupling design for array applications," *IEEE Trans. Antennas Propag*, **vol. 51, no. 10, part 2**, 2936–46, 2003.
11. D. K. Cheng, *Field and Wave Electromagnetics*, 2nd edn., Addison-Wesley, 1992.
12. K. Zhang and D. Li, *Electromagnetic Theory for Microwaves and Optoelectronics*, 2nd edn., Publishing House of Electronics Industry, 2001.
13. L. Brillouin, *Wave Propagation in Periodic Structures*, 2nd edn., Dover Publications, 2003.
14. F. Yang and Y. Rahmat-Samii, "Reflection phase characterizations of the EBG ground plane for low profile wire antenna applications," *IEEE Trans. Antennas Propagat.*, **vol. 51, no. 10**, 2691–703, October 2003.
15. H. Mosallaei and Y. Rahmat-Samii, "Periodic bandgap and effective dielectric materials in electromagnetics: characterization and applications in nanocavities and waveguides," *IEEE Trans. Antennas and Propagation*, **vol. 51, no. 3**, 549–63, 2003.
16. F. Yang, J. Chen, Q. Rui, and A. Elsherbeni, "A simple and efficient FDTD/PBC algorithm for periodic structure analysis," *Radio Sci.*, **vol. 42, no. 4**, RS4004, 2007.
17. P.-S. Kildal, "Artificially soft and hard surfaces in electromagnetics," *IEEE Trans. Antennas Propag.*, **vol. 38, no. 10**, 1537–44, 1990.
18. Z. Ying, P.-S. Kildal, and A.-A. Kishk, "Bandwidth of some artificially soft surfaces," *IEEE APS. Int. Symp. Dig.*, vol. 1, pp. 390–3, Newport, CA, June 1995.
19. E. Lier and P.-S. Kildal, "Soft and hard horn antennas," *IEEE Trans. Antennas Propagat.*, **vol. 36, no. 8**, 1152–7, 1988.
20. A. Aminian, F. Yang, and Y. Rahmat-Samii, "Bandwidth determination for soft and hard ground planes by spectral FDTD: a unified approach in visible and surface wave regions," *IEEE Trans. Antennas Propagat.*, **vol. 53, no. 1**, 18–28, 2005.
21. Y. Rahmat-Samii and H. Mosallaei, "Electromagnetic band-gap structures: classification, characterization and applications," *Proceedings of IEE-ICAP symposium*, pp. 560–4, April 2001.
22. P. K. Kelly, J. G. Maloney, B. L. Shirley, and R. L. Moore, "Photonic bandgap structures of finite thickness: theory and experiment," *IEEE APS Int. Symp. Dig.*, vol. 2, pp. 718–21, June 1994.
23. D. J. Kern, D. H. Werner, A. Monorchio, L. Lanuzza, and M. J. Wilhelm, "The design synthesis of multiband artificial magnetic conductors using high impedance frequency selective surfaces," *IEEE Trans. Antennas Propagat.*, **vol. 53, no. 1, part 1**, 8–17, 2005.
24. G. Goussetis, A. P. Feresidis, and J. C. Vardaxoglou, "FSS printed on grounded dielectric substrates: resonance phenomena, AMC and EBG characteristics," *IEEE APS Int. Symp. Dig.*, vol. 1B, pp. 644–7, 3–8 July 2005.

4 Designs and optimizations of EBG structures

After demonstrating some interesting characteristics of EBG structures, this chapter focuses on how to achieve these characteristics by properly designing the EBG structures. A parametric study on the mushroom-like EBG structure will be presented first. Then two popularly used planar EBG structures, mushroom-like EBG surface and uni-planar EBG surface, are compared with each other to develop some selection guidelines for potential applications. Novel EBG designs such as polarization-dependent EBG (PDEBG), compact spiral EBG, and stack EBG structures will also be studied. Furthermore, the particle swarm optimization (PSO) technique is implemented to design EBG surfaces. At the end of this chapter, some recent research trends are summarized, including space filling curve EBG surfaces, multi-band EBG designs and reconfigurable EBG structures.

4.1 Parametric study of a mushroom-like EBG structure

Electromagnetic properties of an EBG structure are determined by its physical dimensions. For a mushroom-like EBG structure shown in Fig. 4.1, there are four main parameters affecting its performance [1], namely, patch width W, gap width g, substrate thickness h, and substrate permittivity ε_r. In this section, the effects of these parameters are investigated one by one in order to obtain some engineering guidelines for EBG surface designs. Note that the vias' radius r has a trivial effect because it is very thin compared to the operating wavelength.

In this study, the FDTD/PBC technique [2–3] is used to characterize the reflection phase of the EBG structure. Normal incidence is considered here. The dimensions of a reference EBG are listed below:

$$W = 0.12\,\lambda_{12\,\text{GHz}},\ g = 0.02\,\lambda_{12\,\text{GHz}},\ h = 0.04\,\lambda_{12\,\text{GHz}},\ \varepsilon_r = 2.20,\ r = 0.005\,\lambda_{12\,\text{GHz}} \tag{4.1}$$

4.1.1 Patch width effect

Patch width plays an important role in determining the resonant frequency. When studying the effect of the EBG patch width, other parameters such as the gap width, substrate permittivity, and substrate thickness are kept the same as in (4.1). The patch width is changed from $0.04\,\lambda_{12\,\text{GHz}}$ to $0.20\,\lambda_{12\,\text{GHz}}$.

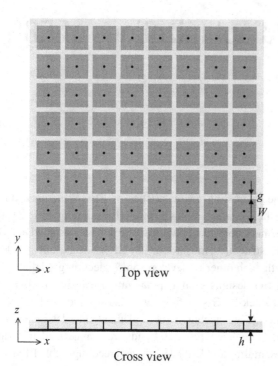

Fig. 4.1 Geometry of a mushroom-like electromagnetic band gap (EBG) structure.

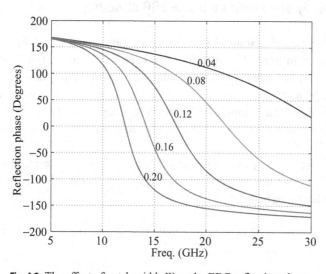

Fig. 4.2 The effect of patch width W on the EBG reflection phase.

Figure 4.2 shows the reflection phases of EBG surfaces with different patch widths. It is observed that when the patch width is increased, the resonant frequency for 0° reflection phase decreases. Furthermore, the slope of the curve becomes steep near the resonance, which indicates a narrow bandwidth. This phenomenon can be explained

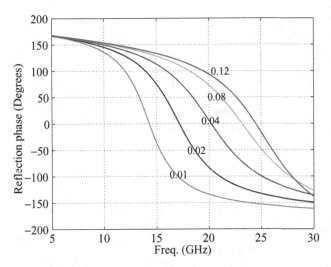

Fig. 4.3 The effect of gap width g on the EBG reflection phase.

using the lumped LC model. According to (3.3), a wider patch width leads to a larger capacitance C. Thus, the frequency reduces and the bandwidth becomes narrow.

4.1.2 Gap width effect

The gap width g controls the coupling between EBG patch units. Hence, variation of the gap width also affects the frequency band of the EBG surface. During this investigation, the patch width, substrate permittivity, and substrate thickness are kept the same as in (4.1). The gap width is increased from $0.01\ \lambda_{12\,\text{GHz}}$ to $0.12\ \lambda_{12\,\text{GHz}}$. When the gap width is increased to $0.12\ \lambda_{12\,\text{GHz}}$, the gap width becomes the same as the patch width.

Figure 4.3 displays the reflection phases of EBG surfaces with different gap widths. It is noticed that the variation in the gap width has the opposite effect to the patch width. When the gap width is increased, the resonant frequency increases. Meanwhile, the slope of the curve becomes flat near the resonance, which indicates a wide bandwidth. According to the lumped LC model, increasing the gap width will decrease the value of the capacitance C. Thus, both the resonant frequency and the bandwidth increases.

4.1.3 Substrate thickness effect

In the previous studies, we noticed that the bandwidth changes in the same way as the resonant frequency. When the frequency decreases, the bandwidth also becomes narrower. From a designer viewpoint, can we decrease the resonant frequency while increasing the bandwidth? This can be achieved by adjusting the substrate thickness h. In the following simulations the patch width, gap width, and substrate permittivity

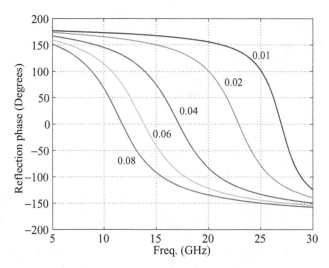

Fig. 4.4 The effect of substrate thickness h on the EBG reflection phase.

are the same as in (4.1). The substrate thickness is changed from 0.01 $\lambda_{12\,GHz}$ to 0.09 $\lambda_{12\,GHz}$. Note that the substrate thickness is always kept small compared to the wavelength because a thin EBG surface is preferred in practical applications.

The reflection phases with different substrate thicknesses are shown in Fig. 4.4. It is observed that if the substrate thickness is increased, the frequency decreases. This is similar to the effect of the patch width. However, the slope of the curve becomes flat and the bandwidth increases with the substrate thickness. This can be also explained from the LC model. When the substrate thickness is increased, the equivalent inductance L increases. Thus, the frequency reduces but the bandwidth increases.

4.1.4 Substrate permittivity effect

Relative permittivity (ε_r) of a substrate, also called the dielectric constant, is another effective parameter used to control the frequency behavior. Some commonly used commercial materials such as RT/duroid substrates and TMM substrates are investigated, as well as air. The EBG structure analyzed here has the same parameters as (4.1), except that the permittivity is changed.

The reflection phases of EBG surfaces with various permittivities are plotted in Fig. 4.5. When air is used as the substrate, the EBG surface has the highest resonant frequency and largest bandwidth. When the permittivity is increased, the resonant frequency decreases, so does the bandwidth. Thus, we can always use a high dielectric constant substrate to reduce the EBG cell size. The price we pay is the narrow bandwidth. This effect is similar to a microstrip patch antenna design.

Although the preceding investigations focus on the reflection phase features of the EBG structure, similar parameter effects are also observed for the surface wave band properties.

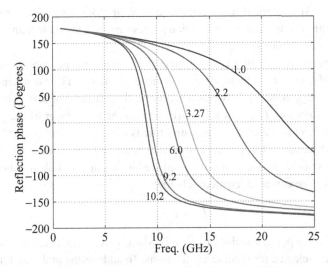

Fig. 4.5 The effect of substrate permittivity ε_r on the EBG reflection phase.

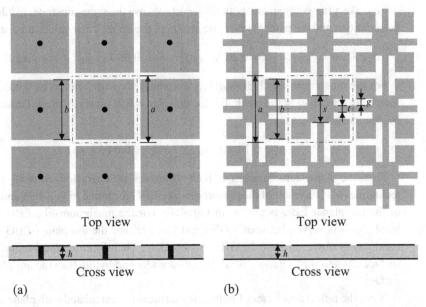

Fig. 4.6 Geometries of (a) a mushroom-like EBG surface and (b) a uni-planar EBG surface. They are designed on the same substrate and have the same periodicity a. Dashed line indicates the periodic boundary condition.

4.2 Comparison of mushroom and uni-planar EBG designs

Numerous EBG surfaces have been proposed and investigated in recent years. Besides the mushroom-like EBG surface, another popular EBG surface is the uni-planar EBG surface [4–5]. Figure 4.6 compares the geometries of the mushroom-like EBG and the uni-planar EBG. An important feature in the uni-planar EBG design is the removal of vertical vias. Thus, it simplifies the fabrication process and is compatible with microwave and millimeter wave circuits.

Similar to the mushroom-like EBG surface, the operational mechanism of the uni-planar EBG surface can be explained by the lumped LC model. The capacitance C also comes from the edge coupling between adjacent patches. Instead of using vertical vias to provide an inductance L, a thin microstrip line on the same layer of the patches is used to connect them together. Thus, this structure is named as a "uni-planar" EBG. To increase the inductance value, the microstrip line needs to be inset into the patch. The main parameters of the metallic pattern design are labeled in Fig. 4.6b and the length l_s of the microstrip line can be calculated as below:

$$l_s = a - s \qquad (4.2)$$

From a user's viewpoint, a viable question is which EBG surface, mushroom or uni-planar, should be selected for specific applications. To address the problem, this section compares the surface wave band gap and the in-phase reflection feature of these two EBG surfaces.

To make a fair comparison, both EBG surfaces use the same substrate. In addition, the periodicity a and the patch width b are also kept the same. Their values are listed below:

$$a = 0.30\ \lambda_{12\ \text{GHz}}, \quad b = 0.26\ \lambda_{12\ \text{GHz}}, \quad h = 0.04\ \lambda_{12\ \text{GHz}}, \quad \varepsilon_r = 2.20 \qquad (4.3)$$

For the mushroom-like EBG surface, a vertical via with a radius of $0.005\ \lambda_{12\ \text{GHz}}$ is located in the center of the patch. For the uni-planar EBG surface, the parameters for the inset microstrip lines are:

$$s = 0.14\ \lambda_{12\ \text{GHz}}, \quad t = g = 0.04\ \lambda_{12\ \text{GHz}} \qquad (4.4)$$

First, the dispersion diagrams of both structures are characterized using the FDTD technique [6]. The result of the mushroom-like EBG is plotted in Fig. 4.7a and the result of the uni-planar EBG is plotted in Fig. 4.7b. For the mushroom-like EBG surface, a band gap is observed between 7 GHz and 11 GHz. For the uni-planar EBG surface, a band gap is observed from 13 GHz to 14.6 GHz. Therefore, the mushroom-like EBG surface has a lower frequency band gap and a wider bandwidth than the uni-planar EBG surface.

Next, the reflection phases of both EBG surfaces are calculated with plane wave incidences. To consider the incidence angle and polarization effects, the normal incidence, the TM incidence at 80 degrees, and the TE incidence at 60 degrees are all simulated using the FDTD technique. The results are presented in Fig. 4.8. For the mushroom-like EBG surface, the resonant frequencies, where the reflection phase is equal to zero degrees, are 7.8 GHz, 11 GHz, and 11.5 GHz for TM80, normal, and TE60 cases, respectively. For the uni-planar EBG surface, the corresponding frequencies are 13 GHz, 13.3 GHz, and 13.7 GHz, respectively. Thus, the mushroom-like EBG surface has lower resonance frequencies than the uni-planar EBG surface. However, the frequency shift in the uni-planar case is smaller than that in the mushroom-like EBG case, which indicates

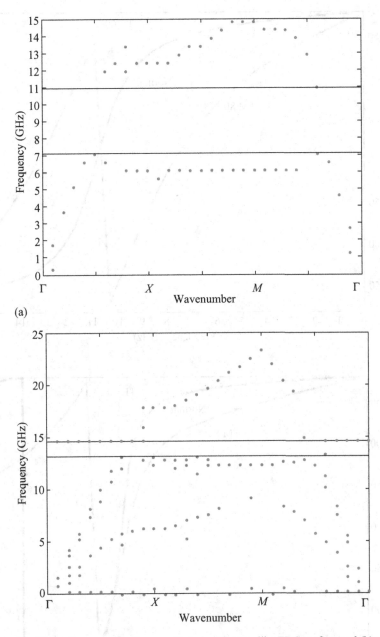

Fig. 4.7 Surface wave band gaps of (a) a mushroom-like EBG surface and (b) a uni-planar EBG surface.

that the uni-planar design is less sensitive to the incident angle and polarization state. Furthermore, we notice that the slopes of the curves in Fig. 4.8a are flatter than those in Fig. 4.8b. It means that the applicable bandwidth of the mushroom-like EBG surface is wider than the uni-planar EBG surface. These observations are consistent with those in Fig. 4.7.

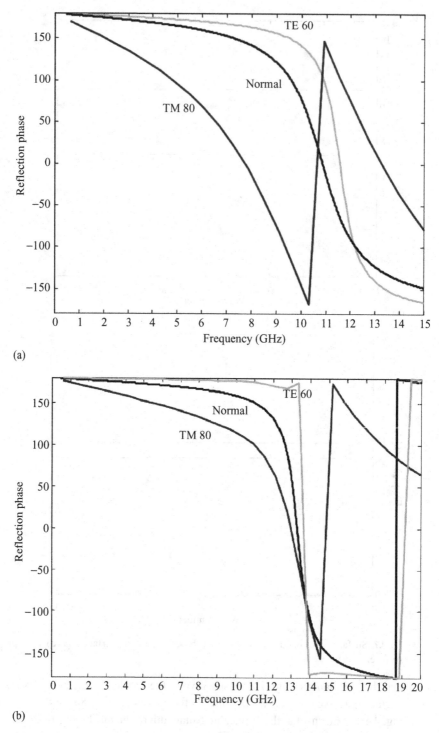

Fig. 4.8 Reflection phases of (a) a mushroom-like EBG surface and (b) a uni-planar EBG surface.

From the above comparisons, we see that the mushroom-like EBG surface and the uni-planar EBG surface each have their own advantages. A brief summary is provided below:

- Advantages of the mushroom-like EBG surface: a lower frequency and a wide bandwidth are obtained. At a given frequency, its size is smaller than the uni-planar EBG design.
- Advantages of the uni-planar EBG surface: it has no vertical vias, which makes the fabrication easier. In addition, it is less sensitive to the incident angle and polarization.

4.3 Polarization-dependent EBG surface designs

The aforementioned EBG surfaces consist of symmetric square units. Thus, the reflection phase for a normally incident plane wave is *independent* of its polarization state. It is an isotropic structure for normal incidence. In this section, novel anisotropic EBG surfaces whose reflection phases are *dependent* on polarization states are investigated. Various polarization-dependent EBG (PDEBG) designs are presented, including rectangular patch EBG surface, slot loaded EBG surface, and EBG surface with offset vias [7–8].

4.3.1 Rectangular patch EBG surface

The traditional mushroom-like EBG structure uses square patch units so that its reflection phase for normal incidence is independent of the polarization. To obtain a polarization-dependent reflection phase, one approach is to replace the square patches by rectangular patches, as shown in Fig. 4.9. For example, the patch length L is 0.24 $\lambda_{3\,GHz}$ and the width W is 0.16 $\lambda_{3\,GHz}$. The gap width is 0.02 $\lambda_{3\,GHz}$ and the vias' radius is 0.0025 $\lambda_{3\,GHz}$. The substrate thickness is 0.04 $\lambda_{3\,GHz}$ and the dielectric constant is 2.20. As a reference, a square patch EBG surface (0.16 $\lambda_{3\,GHz}$ × 0.16 $\lambda_{3\,GHz}$) is also studied.

Due to different values of L and W, the reflection phase of the EBG surface becomes dependent on the x- or y-polarization state of the incident plane wave. Figure 4.10 presents the FDTD simulated reflection phase results of the rectangular patch EBG surface compared to a square patch EBG surface. When the incident plane wave is y-polarized, the rectangular patch EBG surface has the same reflection phase as the square patch EBG surface because the patch widths are the same. For the x-polarized incident plane wave, the patch length L plays a dominant role in determining the reflection phase. Since the length L is longer than the width W, the reflection phase curve shifts down to a lower frequency. It is noticed that near 3 GHz, the EBG surface shows a $-90°$ reflection phase for the x-polarized wave and a $+90°$ reflection phase for the y-polarized wave. Thus, the phase difference between orthogonal polarizations is $180°$.

Designs and optimizations of EBG structures

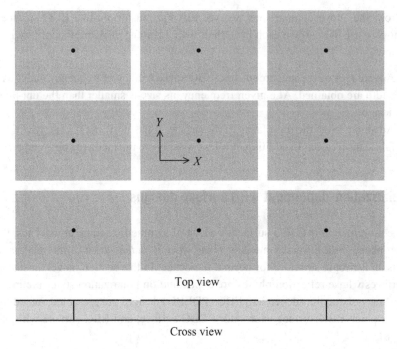

Fig. 4.9 A polarization-dependent EBG surface design using rectangular patches (from [8], © Wiley Inter-Science, 2004).

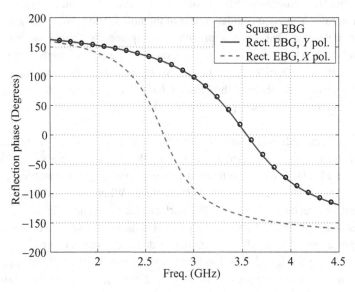

Fig. 4.10 Reflection phases of the rectangular patch EBG surface with comparison to a square patch EBG surface (from [8], © Wiley Inter-Science, 2004).

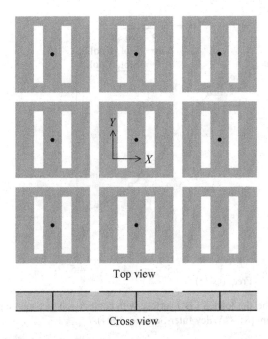

Fig. 4.11 A polarization-dependent EBG surface design using slot loaded patches (from [8], © Wiley Inter-Science, 2004).

4.3.2 Slot loaded EBG surface

Another approach to realize polarization-dependence is to incorporate slots into the patches of an EBG surface, similar to a slot loaded microstrip patch antenna [9–11]. An advantage of this technique is that a square unit can still be used and the patch length does not need to be increased to lower the resonant frequency. Thus, a compact unit size is obtained. As shown in Fig. 4.11, a pair of y-oriented slots is symmetrically incorporated into the patch of a square EBG surface. The dimensions of the square patch EBG are the same as the previous design. The slot length is 0.12 $\lambda_{3\,GHz}$, slot width is 0.02 $\lambda_{3\,GHz}$, and slot separation is 0.06 $\lambda_{3\,GHz}$.

The slots affect the flow of electric currents along the x direction, resulting in a longer current path. Thus, the reflection phase of the x-polarized wave decreases to a low frequency, as depicted in Fig. 4.12. Meanwhile, the reflection phase of the y-polarized wave remains the same as the traditional square patch EBG surface because the slots have little effect on the electric currents flowing along the y direction. The maximum reflection phase difference of this design is 98° obtained at 3.45 GHz. To increase the phase difference between two polarizations, one can increase the length of the slots. As an extreme case, one end of the slots can cut all the way to the edge of the patches.

4.3.3 EBG surface with offset vias

Vias are critical components in a mushroom-like EBG structure. When the vias are offset from the center of the EBG patch, the reflection phase changes and a PDEBG surface

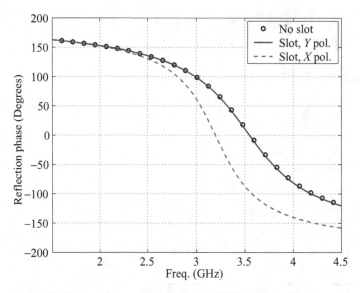

Fig. 4.12 Reflection phases of the slot loaded EBG surface with comparison to a square patch EBG surface without slots (from [8], © Wiley Inter-Science, 2004).

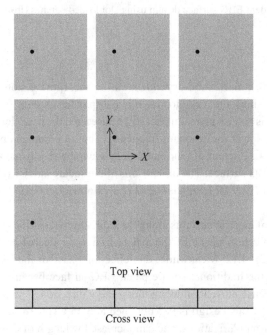

Fig. 4.13 A polarization-dependent EBG surface design with offset vias (from [8], © Wiley Inter-Science, 2004).

can be realized. As shown in Fig. 4.13, the vias are offset along the $-x$ direction while they are still in the center along the y direction. Therefore, the reflection phase for the y-polarized wave remains unchanged and the reflection phase for the x-polarized wave changes with the vias' positions.

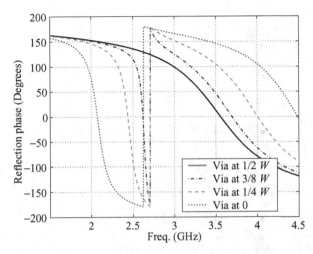

Fig. 4.14 Reflection phases of EBG surfaces with different via positions under x-polarized incident wave (from [8], © Wiley Inter-Science, 2004).

When a via is located in the center of the patch ($\frac{1}{2}W$), only one resonance on the reflection phase is observed, as shown in Fig. 4.14. Once the vias are offset, two resonances appear with one higher than the original frequency and the other lower. Different resonances correspond to the different widths of the left and right parts of the patch with respect to the vias. The left part is narrower and it forms the higher resonance. The right part is wider and resonates at a lower frequency. When the vias move closer to the patch edges, the separation of the two resonances increases because the width difference between two sides of the patch becomes longer.

To further understand these phenomena, an EBG surface with two symmetric vias under each patch is investigated, as shown in Fig. 4.15. The distance from the via's location to the edge of the patch is a quarter of the patch width. Figure 4.16 compares the reflection phase results of the single via case and the double vias case. For the y-polarized incident plane wave, they have the same reflection phase. When the incident plane wave is x-polarized, the double vias case shows only one resonance, which is close to the higher resonance of the single via case. This observation verifies that the resonance feature is mainly determined by the distance between the via's location and the patch boundary.

4.3.4 An example application: PDEBG reflector

Polarization-dependent EBG surfaces can be used in various electromagnetic applications such as a polarization converter [12] or a ground plane for low profile antennas [13]. Here, the implementation of a PDEBG surface as a reflector is presented. It is well known that when a right-hand circularly polarized (RHCP) plane wave impinges onto a perfect electric conductor (PEC) surface, the reflected plane wave becomes left-hand circularly polarized (LHCP). Thus, the polarization sense is reversed. When a PDEBG

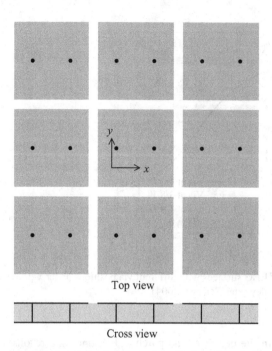

Fig. 4.15 A polarization-dependent EBG surface design with a pair of offset vias on each patch (from [8], © Wiley Inter-Science, 2004).

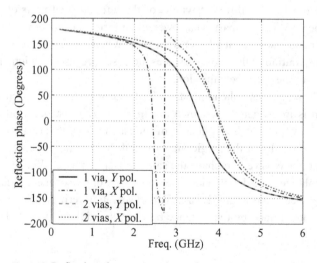

Fig. 4.16 Reflection phase comparison of EBG designs with one offset via and a pair of offset vias under each patch (from [8], © Wiley Inter-Science, 2004).

surface is used as a reflector, the polarization state of the reflected plane wave is determined by the reflection phases of two linearly polarized components. By controlling the reflection phase difference, various polarizations can be realized.

To illustrate the polarization feature of the PDEBG reflector, let's assume that the incident plane wave is right-hand circularly polarized (RHCP) and normally illuminates upon a reflector located in the xy plane,

$$\vec{E}^{\text{i}} = \hat{x} \cdot e^{jkz} + \hat{y} \cdot j \cdot e^{jkz}. \tag{4.5}$$

The time convention is $e^{j\omega t}$. The reflected wave is obtained as follows:

$$\vec{E}^{\text{r}} = \hat{x} \cdot e^{-jkz+j\theta_x} + \hat{y} \cdot j \cdot e^{-jkz+j\theta_y} \tag{4.6}$$

where θ_x is the reflection phase for the x-polarized wave and θ_y is the reflection phase for the y-polarized wave. Equation (4.6) can be organized in terms of two circularly polarized components as:

$$\vec{E}^{\text{r}} = e^{-jkz} \cdot e^{j\theta_x} \cdot \left[\hat{E}_{\text{L}} \cdot \left(\frac{1+e^{j(\theta_y-\theta_x)}}{\sqrt{2}}\right) + \hat{E}_{\text{R}} \left(\frac{1-e^{j(\theta_y-\theta_x)}}{\sqrt{2}}\right)\right]. \tag{4.7}$$

$$\hat{E}_{\text{L}} = \frac{\hat{x}+\hat{y}\cdot j}{\sqrt{2}}, \tag{4.8}$$

$$\hat{E}_{\text{R}} = \frac{\hat{x}-\hat{y}\cdot j}{\sqrt{2}}. \tag{4.9}$$

$\hat{E}_{\text{L}}$ is a unit vector that is left-hand circularly polarized (LHCP) and $\hat{E}_{\text{R}}$ is a unit vector that is right-hand circularly polarized (RHCP).

For a traditional reflector such as a PEC reflector, $\theta_x = \theta_y$. So the coefficient of the RHCP component is zero, and the reflected wave is purely LHCP. The same phenomenon happens for a PMC reflector or a conventional EBG surface whose reflection phase is independent of x- or y-polarizations. Thus, the RHCP plane wave changes to LHCP plane wave after reflection from these surfaces.

To maintain the same polarization sense, the PDEBG is utilized as the reflector. The reflection phases of x- and y-polarized waves are different, and their difference changes with frequency. At a certain frequency, the phase difference is 180°. Substituting this phase difference into (4.7), it can be found that the coefficient of the LHCP component becomes zero. As a result, the reflected wave remains RHCP.

To demonstrate this idea, the rectangular patch EBG surface is used as an example. The phase difference $(\theta_y - \theta_x)$ versus frequency is plotted in Fig. 4.17a. When the frequency is low, the phase difference is very small. The phase difference increases with the frequency increasing. When the frequency is 2.90 GHz, the phase difference is 180°. The phase difference reaches its maximum (193°) at 3.09 GHz and after that it decreases with frequency.

The axial ratio (AR) of the reflected wave is plotted in Fig. 4.17b. It is calculated from the following equation [14]:

$$\text{AR} = -\frac{\|E_{\text{R}}\| + \|E_{\text{L}}\|}{\|E_{\text{R}}\| - \|E_{\text{L}}\|} = -\frac{\left\|\frac{1-e^{j(\theta_y-\theta_x)}}{\sqrt{2}}\right\| + \left\|\frac{1+e^{j(\theta_y-\theta_x)}}{\sqrt{2}}\right\|}{\left\|\frac{1-e^{j(\theta_y-\theta_x)}}{\sqrt{2}}\right\| - \left\|\frac{1+e^{j(\theta_y-\theta_x)}}{\sqrt{2}}\right\|} \tag{4.10}$$

It is observed that at 2.90 GHz and 3.30 GHz, the axial ratio is equal to −1 and a purely RHCP reflected wave is obtained. The frequency band inside which the axial ratio of the RHCP reflected wave is below 3 dB ranges from 2.81 GHz to 3.43 GHz (19.9%).

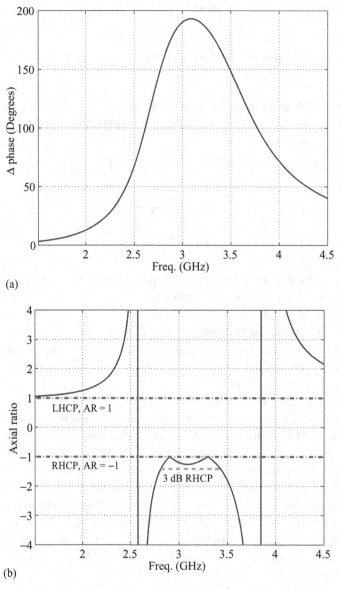

Fig. 4.17 (a) Reflection phase difference between different polarizations of the rectangular PDEBG surface shown in Fig. 4.9. (b) Axial ratio of the reflected plane wave from the PDEBG reflector (from [8], © Wiley Inter-Science, 2004).

It is noted that the reflected wave can also be linearly polarized (LP) or LHCP, depending on the frequency and reflection phase difference. For example, at 2.58 GHz, the reflection phase difference is 90°. Thus, the reflected plane wave is linearly polarized with an axial ratio equal to infinity. The field direction is $\hat{x} + \hat{y}$ according to (4.7). Furthermore, if one reverses the roles of the incident and reflected waves, it can be inferred from Fig. 4.17b that no matter what the polarization of the incident wave is, the PDEBG

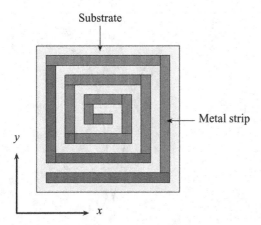

Fig. 4.18 Unit cell geometry of a single spiral EBG surface.

reflector can change it to a purely RHCP plane wave by properly designing the phase difference between two x-polarized and y-polarized waves.

4.4 Compact spiral EBG designs

An important topic in EBG research is to miniaturize the cell size of the EBG unit. When the cell size is small compared to wavelength, the surface can be considered as a homogeneous medium. A wide variety of EBG surfaces have been reported such as Hilbert-curve design [15], Jerusalem crosses [16], dipole and slot arrays [17]. In particular, the Hilbert-curve inclusion focuses on the compactness of the surface. As the number of iterations increases, the equivalent inductance increases, resulting in a lower resonant frequency. Using a similar idea, a larger equivalent inductance can also be realized with an increasing number of spiral turns [18]. However, as revealed in this section, if unit geometry of a spiral EBG is not symmetric with respect to the polarizations of the incident waves, the EBG surface generates a high level of cross polarization. Thus, the behavior of the reflection phase may not be applicable in a designated frequency band of operation. Several typical printed spiral geometries are investigated in this section and their reflection properties are reported.

4.4.1 Single spiral design

According to the lumped LC model of the EBG surface, the resonance frequency can be reduced by increasing the inductance value. In light of this, a single spiral is placed on top of a grounded substrate to replace the conventional square patch, as shown in Fig. 4.18. The thickness of the substrate is $0.04\ \lambda_{12\ GHz}$ and the relative permittivity of the substrate is 2.2. The periodicity of the unit cell is $0.15\ \lambda_{12\ GHz}$. The width of the spiral is $0.01\ \lambda_{12\ GHz}$ and the gap between metal strips is also $0.01\ \lambda_{12\ GHz}$. Periodic boundary conditions (PBC) are positioned on four sides of the unit cell to model the effect of periodic replication in an infinite array structure.

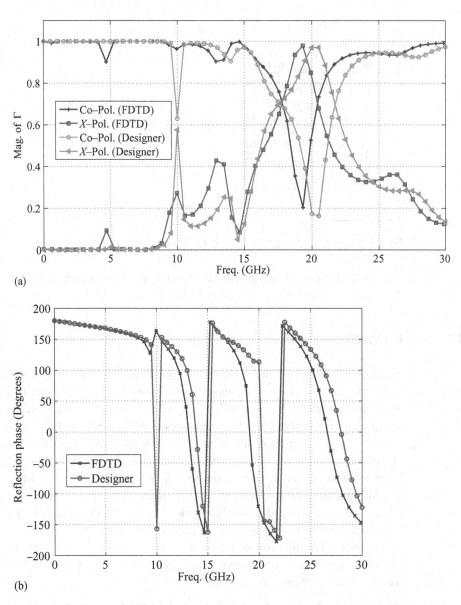

Fig. 4.19 Reflection results of the single spiral EBG surface: (a) reflection magnitudes and (b) reflection phases for co-polarized fields (from [21], © The Electromagnetics Academy, 2007).

A y-polarized plane wave is normally incident on the EBG surface. Both the FDTD technique [19] and Ansoft Designer [20] are used to characterize the EBG performance. Figure 4.19a shows the magnitudes of the reflection coefficients for both the co-polarized field and cross-polarized field. A large cross polarization is observed in the frequency band from 10 GHz to 30 GHz because of the *asymmetric geometry*. Four magnitude resonances can be found at 9.6 GHz, 13.1 GHz, 19.1 GHz, and 26.6 GHz. In Fig. 4.19b,

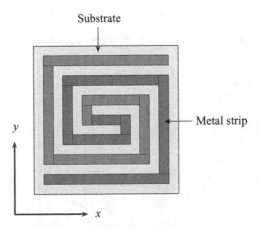

Fig. 4.20 Unit geometry of a double spiral EBG surface.

the 0° phase exists at the same frequencies as the magnitude resonances. Although the first resonant frequency is lower than that of a square patch EBG, the resulting high cross polarization is not acceptable in many applications.

4.4.2 Double spiral design

To reduce the cross polarization, a double spiral geometry is investigated, as shown in Fig. 4.20. The dielectric substrate and periodicity of the unit cell remain the same. The width of spiral is also 0.01 λ_{12GHz}. Compared to the single spiral EBG in Fig. 4.19, it is more symmetric. If the geometry is rotated 180 degrees, it will recover itself.

The simulated reflection results from the FDTD technique and Ansoft Designer are shown in Fig. 4.21. It is noticed that the overall cross polarization has been reduced as compared to Fig. 4.19. The resonances of both reflection magnitude and phase can be found at 13.1 GHz and 27.4 GHz. From the simulated results, the resonances in the reflection magnitudes, which exhibit cross polarization, again match to the phase resonances.

4.4.3 Four-arm spiral design

To completely eliminate the cross polarization, a four-arm spiral is explored, as shown in Fig. 4.22 [21]. Four spiral branches, each with a 0.01 $\lambda_{12\ GHz}$ width, split from the center and rotate outwards. For this specific design, if this unit cell is rotated 90 degrees, it can exactly recover itself. Thus, this geometry is symmetric not only in +/−x *or* +/−y directions, but also in x *and* y directions. The scattering responses to the x- and y-polarized fields are identical. As a result, the cross-polarized components can be canceled.

Figure 4.23 presents the FDTD and Designer simulated results, where cross-polarized fields are zero. In the Ansoft Designer data, a null in the co-polarized field is observed

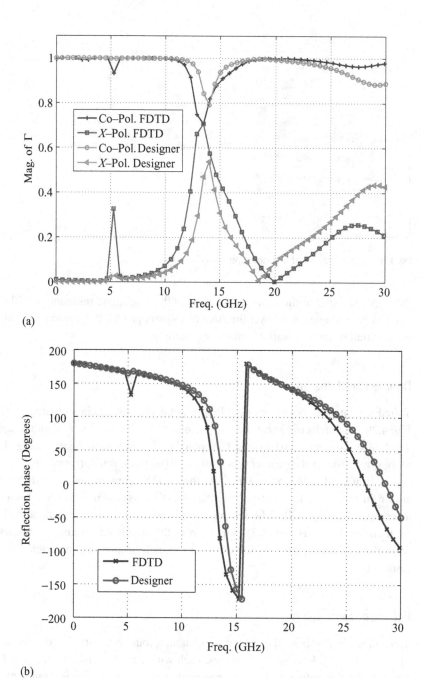

Fig. 4.21 Reflection results of the double spiral EBG surface: (a) reflection magnitudes and (b) reflection phases for co-polarized fields (from [21], © The Electromagnetics Academy, 2007).

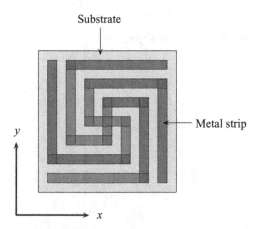

Fig. 4.22 Unit cell geometry of a four-arm spiral EBG surface.

near 9.2 GHz, which may be caused by a difficulty in singularity treatment near the resonance. The resonances of the reflection phase can be found at 9.2 GHz and 21.2 GHz in Fig. 4.23b. As a reference, a conventional square patch EBG surface with the same substrate and same periodicity is simulated. The result is also plotted in Fig. 4.23b and the resonant frequency is noticed to be 16.2 GHz. Thus, the first resonant frequency of the spiral EBG surface is 43.2% lower than the square patch EBG.

In summary, the compactness of unit cell geometry is achieved using the four-arm spiral geometry without generating the cross polarization field. The significant reduction in size for a single element leads to a very attractive design feature for applications based on arrays composed of such an element. The resonant frequency can be further decreased as the number of spiral turns increases.

4.5   Dual layer EBG designs

Besides the single layer EBG surfaces, this section investigates the performance of multi-layer stacked EBG designs, as shown in Fig. 4.24. To illustrate their properties, EBG structures with a total thickness of 0.04 $\lambda_{12\ GHz}$ are simulated using the FDTD technique. The dielectric constant is set to 1, which could provide a broad bandwidth. The size of square patch is 0.12 $\lambda_{12\ GHz}$ and the gap width is 0.02 $\lambda_{12\ GHz}$. The height of the top layer is 0.04 $\lambda_{12\ GHz}$ while the height of the middle layer is 0.02 $\lambda_{12\ GHz}$. The top layer and bottom layer are aligned in two different arrangements. Figure 4.24a depicts the geometry of case 1, where the top layer is shifted half a cell size in horizontal and vertical directions with respect to the bottom layer. Figure 4.24b shows the geometry of case 2, where the upper patches are put exactly on top of the bottom patches. These two layers share the same periodic boundary.

The FDTD simulated reflection phase results are plotted in Fig. 4.25. As a reference, a single layer EBG design is also simulated for comparison. It has the same dimensions as the geometries in Fig. 4.24 except that the bottom patch layer is removed. It is noticed that

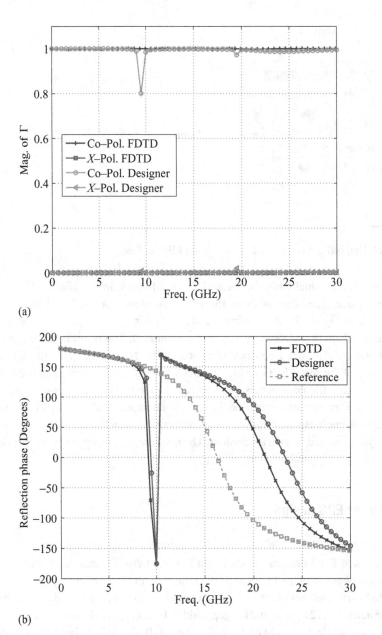

Fig. 4.23 Reflection results of the four-arm spiral EBG surface: (a) reflection magnitudes and (b) reflection phases for co-polarized fields (from [21], © The Electromagnetics Academy, 2007).

the multi-layer designs have lower resonant frequency than that of the single layer EBG surface. The resonant frequency for case 1 is 17.1 GHz and the resonant frequency for case 2 is 21.0 GHz whereas the resonant frequency of the single layer EBG is 22.1 GHz. This can be explained from the lumped LC model. The additional layer provided more coupling between adjacent patches [22]. Thus, the value of capacitance is increased and

4.5 Dual layer EBG designs

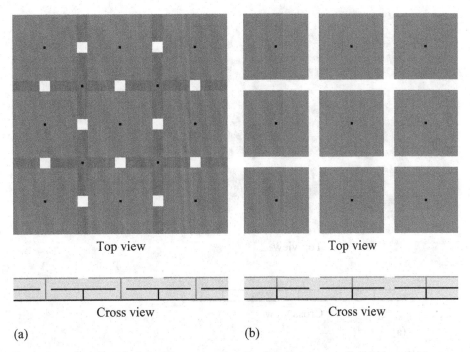

Fig. 4.24 Stacked EBG designs: (a) case 1: unit cells of two layers are offset from each other, (b) case 2: unit cells of two layers share the same periodic boundary.

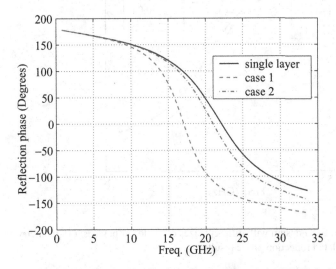

Fig. 4.25 FDTD simulated reflection phases of the stacked EBG designs in Fig. 4.24.

the resonant frequency is decreased. It is also observed that case 1 has lower resonant frequency than case 2. The reason is that the offset dual layer design provides more overlapping areas between adjacent patches. Therefore, the capacitance is increased more, resulting in a lower resonant frequency.

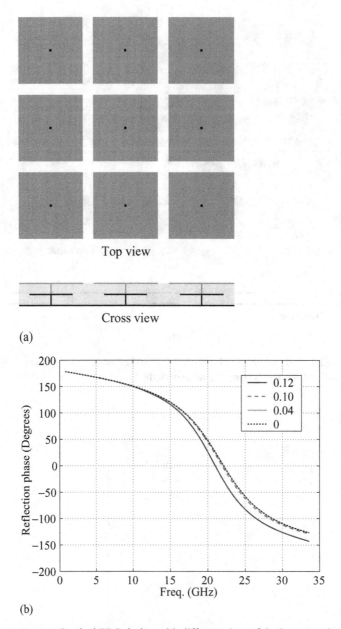

Fig. 4.26 Stacked EBG design with different sizes of the lower patches: (a) geometry and (b) FDTD simulated reflection phase results.

To further understand the performance mechanisms, different patch sizes are studied. In Fig. 4.26a, the patch size of the top layer remains the same while the size of the bottom patch is varied. The FDTD simulated results are shown in Fig. 4.26b. The resonant frequency is increased from 21.0 GHz to 22.1 GHz when the bottom patch size is reduced from 0.12 $\lambda_{12\,GHz}$ to 0. A slight frequency increase is observed.

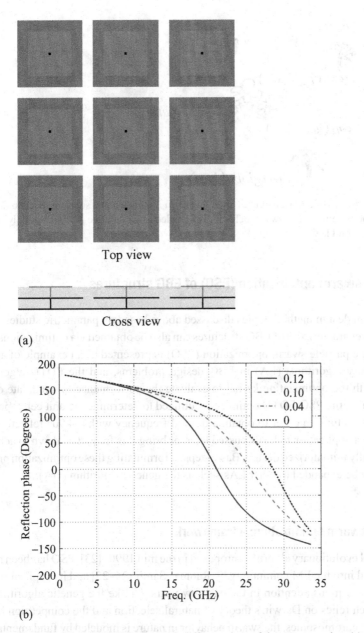

Fig. 4.27 Stacked EBG design with different sizes of the upper patches (a) geometry and (b) FDTD simulated reflection phase results.

In Fig. 4.27a, the patch size of the bottom layer remains the same while the size of the top patch is varied. The FDTD simulated results are shown in Fig. 4.27b. As the patch size is reduced from $0.12 \lambda_{12 \text{ GHz}}$ to 0, the resonant frequency is increased from 21.0 GHz to 28.6 GHz. The frequency change is more noticeable than that in Fig. 4.26, which indicates that the coupling on the top layer is stronger than the coupling in the bottom layer.

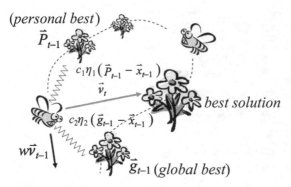

Fig. 4.28 A swarm of ambivalent bees are flying in an N-dimensional solution space to find the highest flower density. The swarm behavior is modeled as an iterative process by applying Newton's second law.

4.6 Particle swarm optimization (PSO) of EBG structures

Aside from design methodologies discussed above based on parametric studies, desired reflection characteristics of EBG structures can also be obtained via optimizations. In this section, the particle swarm optimization (PSO) is presented as an example of applying evolutionary algorithms (EA) in EBG design problems, and the FDTD algorithm is linked with the optimization kernel to evaluate the performance of candidate designs. In particular, the PSO/FDTD optimizer is utilized to determine the unit cell topology of an EBG structure, in order to obtain a desired frequency with a $+90°$ reflection phase for antenna application to be discussed later, or bring this frequency down to achieve an equivalently miniaturized design. Basic steps of formulating these optimization problems can easily be expanded to other EAs such as the genetic algorithm (GA).

4.6.1 Particle swarm optimization: a framework

As a novel evolutionary algorithm proposed in the mid 1990s [23], PSO has been recently introduced into the EM community by Rahmat-Samii [24–25] and its applications have received enormous attention in the past few years. Unlike the genetic algorithm (GA) [26], which relies on Darwin's theory of natural selection and the competition between individual chromosomes, the swarm behavior in nature is modeled by fundamental Newtonian mechanics for optimization purposes. This co-operative scheme manifests PSO's concise formulation, the ease in implementation, and many distinct features in different types of optimizations.

Figure 4.28 depicts a swarm of ambivalent bees flying in a field. The goal of each agent in the swarm is to find a location with the highest flower density. In EM optimization problems, each location in the field is mapped into a candidate design, and the "flower density" corresponds to the performance of the design, which is represented by a fitness function. If the velocities and positions at the tth iteration of all M agents are represented by $M \times N$ matrices $\mathbf{V}_t$ and $\mathbf{X}_t$, the swarm behavior can be modeled as an iterative process by applying Newton's second law [27]:

$$\mathbf{X}_t = \mathbf{X}_{t-1} + \mathbf{V}_t; \qquad (4.11)$$

$$\mathbf{V}_t = w\mathbf{V}_{t-1} + c_1\eta_1\left(\mathbf{P}_{t-1} - \mathbf{X}_{t-1}\right) + c_2\eta_2\left(\mathbf{G}_{t-1} - \mathbf{X}_{t-1}\right). \qquad (4.12)$$

Here $\mathbf{P}_{t-1}$ and $\mathbf{G}_{t-1}$ are matrices that store the best solution that has been visited by each agent (the personal best) and by the entire swarm (the global best), respectively. w is an inertia weight that varies from 0.9 to 0.4. The spring-like attractions are randomized by coefficients $c_1\eta_1$ and $c_2\eta_2$. Typically $c_1 = c_2 = 2.0$ and η_1, η_2 are random numbers with uniform distributions in $(0, 1)$. Equation (4.12) clearly shows that each agent utilizes its own knowledge and communicates with its counterparts during the optimization. Since the position/velocity updating is the only operator required by PSO, an extensive parameter study with respect to multiple operators (such as the crossover, selections, and mutation in GA) is not needed in the PSO algorithm.

Equations (4.11) and (4.12) are both formulated by assuming all variables are real-valued, which makes PSO inherently a good tool for geometrical parameter optimizations in EBG designs. For instance, in [28], the real-number PSO is applied to determine the patch size, the gap width, and the substrate thickness of a mushroom EBG structure. However, the examples in this section are focused on generalized EBG design problems, in which there is no a priori knowledge of the unit cell configuration. In this scenario, the entire design space is typically discretized and represented by a number of pixels. The metallic/dielectric status of each pixel is related to the polarity (0 or 1) of each bit in a binary string. In binary PSO, the velocity of each agent is still updated by following (4.12) while the sigmoid limiting transformation is imposed into the position updating in (4.11) to enable the vector summation in a binary solution space. The readers may refer to [27] for detailed implementations and functional testing results of both real-number and binary PSO algorithms.

With the basic equations formulated, a flowchart of the PSO algorithm is shown in Fig. 4.29. In a binary optimization, the entire process starts by randomly assigning binary strings to all agents, and each agent is evolved within the solution space by updating its velocity and position till the stopping criteria is met. In EBG design problems in this section, the fitness value of each candidate design is related to its reflection characteristics, which is simulated by the FDTD algorithm. For each specific optimization problem, an interface needs to be programmed between the optimization kernel and the external simulator, in order to perform necessary data transferring such as writing FDTD configuration files. In order to reduce the extreme computational cost introduced by full-wave analysis, the platform can be implemented on parallel clusters by executing the fitness evaluations of different agents simultaneously [29–30].

4.6.2 Optimization for a desired frequency with a +90° reflection phase

As a starting point, let us consider the EBG design problem illustrated in Fig. 4.30. The goal of the design is to determine the unit cell topology to obtain a +90° normal reflection

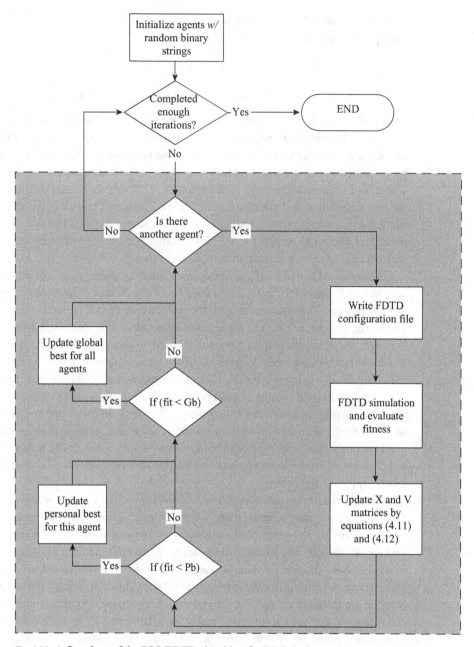

Fig. 4.29 A flowchart of the PSO/FDTD algorithm for EBG design problems.

phase at an arbitrarily selected frequency of $f_0 = 6$ GHz. The reflection characteristics are desired to be independent on all polarization states. The size of each unit cell is fixed to be 7.5 mm × 7.5 mm, and the EBG is designed on a 3 mm-thick RT/Duroid 5800 ($\varepsilon_r = 2.2$) substrate. In order to maintain both the geometrical flexibility and the fabrication feasibility, the design space is discretized into 10 × 10 pixels, which

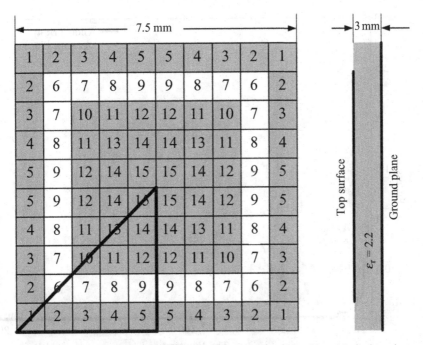

Fig. 4.30 A 7.5 mm × 7.5 mm EBG unit cell is discretized into 10 × 10 pixels and represented by a 15-bit binary string. A four-fold symmetry is imposed into the design template to maintain the polarization independence.

provides a resolution of 0.75 × 0.75 mm. It is not necessary to impose vias in the design template since it does not affect the reflection characteristics for normal incidence.

Theoretically such a unit cell needs to be represented by 100 pixels; however, in order to maintain the polarization independence, a four-fold symmetry must be imposed as shown in Fig. 4.30 [30]. As a result, the dimensionality of solution space (the number of bits required to represent the design) is reduced to fifteen, and the metallic/dielectric status of each pixel is related to the polarity of each bit, B_i ($i = 1, 2, \ldots, 15$), by:

$$\text{pixel } i \text{ is } \begin{cases} \text{metallic} & B_i = 1 \\ \text{dielectric} & B_i = 0 \end{cases} \quad (4.13)$$

For instance, the metallic ring illustrated in Fig. 4.30 is represented by:

$$\{B_i\} = \{0\,0\,0\,0\,0\,1\,1\,1\,1\,0\,0\,0\,0\,0\,0\}. \quad (4.14)$$

Defining the fitness function is usually the most critical part in formulating an optimization problem. In this example, since the PSO algorithm is a minimizer by default, the fitness function is defined as:

$$\text{fitness} = |f_{\Gamma = 90°} - f_0|, \quad (4.15)$$

where $f_{\Gamma = 90°}$ is the simulated frequency with a $+90°$ reflection phase. In other words, the optimizer tries to manipulate the S-curve of candidate design, in order to locate the

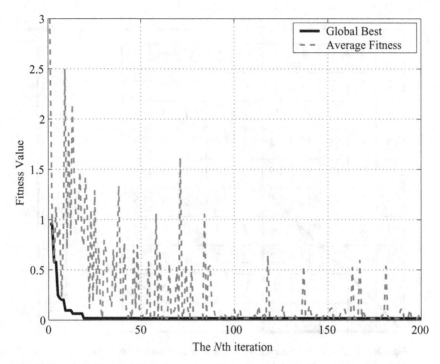

Fig. 4.31 The global best and average fitness value of the optimization using the fitness function defined by (4.15).

$\Gamma = 90°$ frequency as close to the desired $f_0 = 6$ GHz as possible. The 15-dimensional optimization problem is solved using 10 agents for 200 iterations. At each iteration, the topologies of all candidate designs are checked before FDTD simulation. For fabrication purposes, the diagonal connections between pixels are prohibited, and those designs with diagonal connections are directly assigned a very bad fitness value of 1×10^5 without being simulated. Moreover, since the optimization is executed in a binary space, some candidate designs will be encountered more than once. In order to reduce the computational time, these designs are not repeatedly simulated and their fitness values are assigned according to simulation results in previous iterations.

Figure 4.31 plots convergence curves of the optimization. The best design is observed at the 20th iteration, and all agents are well converged to this global optimum towards the end of optimization, as indicated by the equal average fitness value and the global best at the 200th iteration. The simulated S-curve of optimal design is shown in Fig. 4.32, and a $+90°$ reflection phase is observed at 5.99 GHz. The optimized unit cell is shown by the inset of Fig. 4.32, with metallic parts plotted in white and dielectric parts plotted in dark gray. The binary string that represents the optimal design is:

$$\{B_i\} = \{1\ 0\ 0\ 0\ 0\ 0\ 1\ 0\ 0\ 1\ 1\ 1\ 1\ 0\ 0\}. \tag{4.16}$$

4.6 Particle swarm optimization of EBG structures

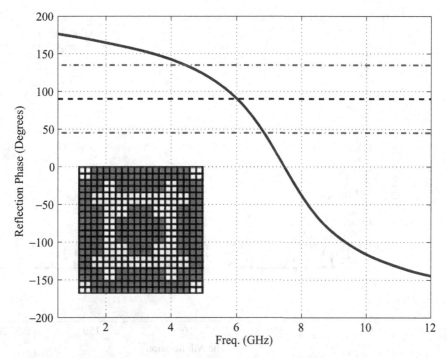

Fig. 4.32 Simulated S-curve of the optimal design with a +90° reflection phase observed at 5.99 GHz. The unit cell topology of optimal design is shown by the inset, with metallic parts plotted in white and dielectric parts in dark gray.

It is worthwhile to mention that each pixel in the design template is modeled by four FDTD grids in full-wave simulation, which guarantees the numerical accuracy for simulating detailed geometries such as narrow metallic strips.

4.6.3 Optimization for a miniaturized EBG structure

Using the same template as shown in Fig. 4.30, we are now re-directing ourselves to the design of a miniaturized EBG structure. In fact, with a fixed unit cell dimension of 7.5 mm × 7.5 mm, the EBG miniaturization problem resembles obtaining the lowest $\Gamma = 90°$ frequency. Therefore it is quite straightforward to define the fitness function as:

$$\text{fitness} = f_{\Gamma = 90°}. \tag{4.17}$$

Figure 4.33 plots the convergence curves using similar optimization setups with 10 agents for 200 iterations. The best design appears at the 25th iteration, which is represented by a binary string of

$$\{B_i\} = \{0\ 0\ 0\ 0\ 0\ 1\ 1\ 1\ 1\ 0\ 0\ 0\ 1\ 1\ 1\} \tag{4.18}$$

As observed from the simulated S-curve shown in Fig. 4.34, the reflection phase is +90° at 5.41 GHz. The optimized unit cell topology is shown in the same figure.

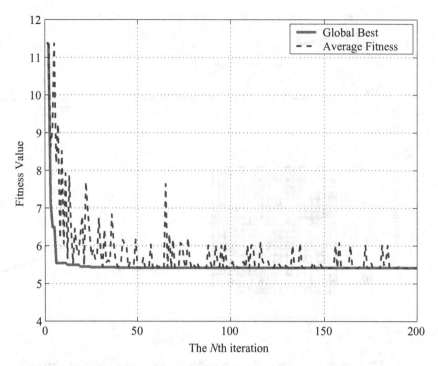

Fig. 4.33 The global best and average fitness value of the optimization using fitness function defined by (4.17).

Although there is not a rigorous proof of the design's optimality, a comparative study has demonstrated that the optimized $\Gamma = 90°$ frequency of 5.41 GHz is lower than, at least, the $\Gamma = 90°$ frequencies of many representative EBG structures. For instance, as shown in Fig. 4.34, the square EBG has a $\Gamma = 90°$ frequency at 5.7 GHz and the UCEBG proposed in [4] has a much higher $\Gamma = 90°$ frequency at 8.8 GHz. It is also noteworthy that for a "connected" version of optimal design (by changing B_5 from 0 to 1, which corresponds to a 2-pixel-wide metallic strip between adjacent unit cells), the simulated S-curve is almost flat with a constant value of $\Gamma = 180°$ in the interested frequency range. This agrees with our physical intuition that an EBG consisting of connected unit cells performs more like a PEC at relatively low frequencies. The optimal design has been successfully applied to the design of a low-profile wire antenna [30], in which the impedance matching of the antenna is significantly improved at a frequency close to 5.41 GHz.

4.6.4 General steps of EBG optimization problems using PSO

As a summary of the examples presented above, the general steps of applying PSO to EBG designs are as follows:

- Discretize the design space and impose necessary geometrical constraints such as intrinsic symmetrical condition, diagonal pixel connection, etc.

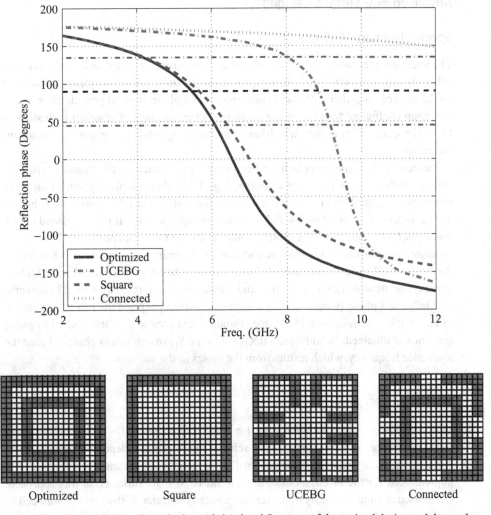

Fig. 4.34 Unit cell topologies and simulated S-curves of the optimal design and three other representative EBG designs.

- Build up the interface between PSO kernel and the EM simulator for necessary data transferring.
- Define the fitness function.
- Select appropriate numbers of agents and iterations. Generally the number of agents should be at least comparable to the dimensionality of solution space.
- Check the convergence of optimization and the performance of optimal design.

Finally, it should be pointed out that these examples are only focused on the reflection characteristics of EBG structures. For applications that require other EBG features such as surface wave suppression, the optimal designs discussed above need to be further investigated by other numerical techniques such as spectral FDTD [31].

4.7 Advanced EBG surface designs

4.7.1 Space filling curve EBG designs

The interesting features of EBG can also be achieved using resonant inclusions on top of a thin grounded substrate [32]. Near the resonance frequency, strong currents are induced on the surface. Together with the conducting ground plane, it could provide an in-phase reflection coefficient for plane wave incidence. One approach of resonant inclusions is to use the space filling curves, which have an electrically small footprint at their resonant frequency.

Various space filling curves have been investigated, such as the Peano curve [33] and the Hilbert curve [34]. For example, Fig. 4.35a shows a photograph of an EBG prototype. The third-order Hilbert curves were fabricated on a 1.575-mm FR-4 substrate with a dielectric constant of 4.4 and loss tangent equal to 0.02. It is then placed on a 5 mm thick foam spacer above a ground plane. Thus, the total thickness is 6.575 mm. For measurement purposes, it is designed so that the resonant frequency would fall within the range of a WR-430 waveguide (1.7 to 2.6 GHz). The final dimensions for the Hilbert curve elements were 12 mm × 12 mm and the surface was formed by 19 × 23 elements. The reflection phase is measured in a waveguide setup [32] and the results are shown in Fig. 4.35b. The reflection phase goes through zero around 2.5 GHz and an in-phase reflection is obtained. In addition, a decrease in the S_{11} magnitude is observed near the resonance frequency, which results from the losses in the substrate.

4.7.2 Multi-band EBG surface designs

Multi-band operation is an interesting topic in EBG research. Using the higher-order mode resonance, an EBG surface may achieve the in-phase reflection at several operating frequencies specified by users. In [35], a genetic algorithm (GA) technique is implemented for the synthesis of optimal multi-band EBG surfaces. In this approach, GA is used to simultaneously optimize the geometry and size of the periodic unit cell as well as the thickness and dielectric constant of the substrate material.

As an example, a dual band EBG surface operating at GPS frequency of 1.575 GHz and cellular frequency of 1.96 GHz is optimized. The final design has a unit cell size of 2.96 cm × 2.96 cm. It is built on a 2.93 mm thick substrate with a dielectric constant of 13. This structure was also fabricated and tested. A photograph of the fabricated EBG surface is shown in Fig. 4.36a, and the simulated versus measured data for this design are plotted in Fig. 4.36b. It is noticed that the targeted resonant frequencies were successfully achieved, with a percentage bandwidth of 4.43% at 1.575 GHz and 2.2% at 1.96 GHz. The phase plot also shows very good agreement between the simulation and the measurements of the actual fabricated surface.

4.7.3 Tunable EBG surface designs

Based on the lumped LC model, the resonance frequency and the reflection phase of an EBG surface can be tuned by changing the effective capacitance, inductance, or both.

(a)

(b)

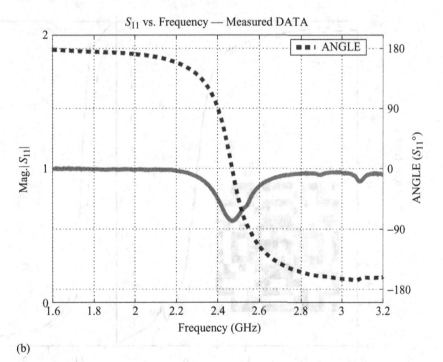

Fig. 4.35 A Hilbert curve based EBG surface: (a) a photograph of a fabricated EBG surface and (b) measured magnitude and phase of the reflection coefficient (from [32], © John Wiley & Sons, 2006).

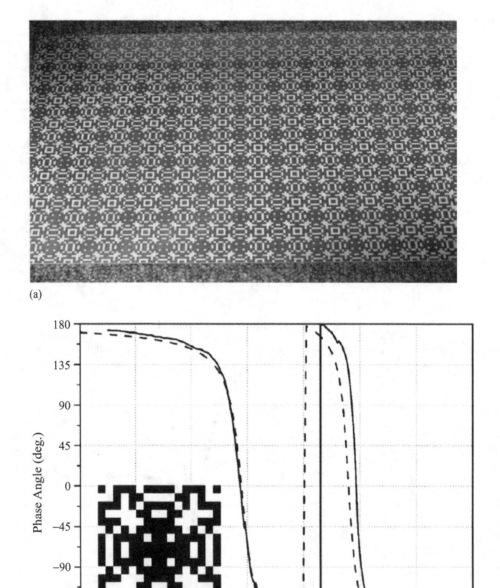

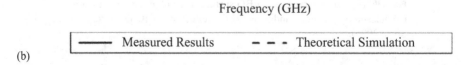

Fig. 4.36 A dual band GA optimized EBG surface operating at GPS and cellular phone bands: (a) a photograph of a fabricated sample and (b) simulated versus measured reflection phase results (from [35], © IEEE, 2005).

4.7 Advanced EBG surface designs

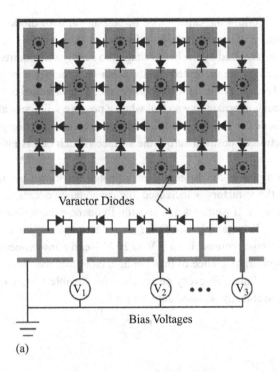

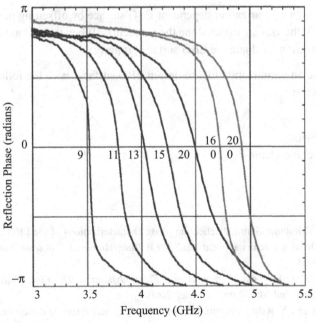

Fig. 4.37 A tunable EBG surface with varactor diodes: (a) schematic plot and (b) measured reflection phase results (from [38], © John Wiley & Sons, 2006).

An attractive concept is to control the capacitance by adding varactor diodes [36–37]. Thus, an electrically tunable EBG surface can be achieved, as shown in Fig. 4.37a [38]. The capacitance value is determined by the bias voltage. To supply the required voltage to the varactors, half of the vias are grounded and the other half vias are connected to a voltage control network through a hole on the ground plane. The varactors are oriented in opposite directions in each alternate row so that when a positive voltage is applied to the control lines, all the diodes are reverse biased. Furthermore, the reflection phase can be programmed as a function of position across the surface if each cell is individually addressed.

The measured reflection phases for different bias voltages are shown in Fig. 4.37b. When the voltage across the varactors is increased, the capacitance decreases, and the resonance frequency increases. If we focus at a specific frequency, the reflection phase increases with bias voltage. For example at 4 GHz, the reflection phase increases from $-\pi$ to π when the bias voltage increase from 9 V to 20 V. Nearly any reflection phase can be realized by the appropriate choice of bias voltage. Note that the two right-most curves are for alternating voltages on every other row. This tunable EBG surface has been used as a simple electronically scanned antenna [39].

4.8 Projects

Project 4.1 Design a polarization-dependent EBG surface by offsetting the vias' position. In Fig. 4.13, the vias are offset along the x direction. When the vias are offset along the diagonal direction, evaluate the EBG surface property.

Project 4.2 Use an optimization tool to design EBG surfaces with the following properties, respectively:

(a) compact unit size;
(b) broad bandwidth;
(c) multiple operation bands.

References

1. F. Yang and Y. Rahmat-Samii, "Reflection phase characterizations of the EBG ground plane for low profile wire antenna applications," *IEEE Trans. Antennas Propagat.*, **vol. 51, no. 10,** 2691–703, 2003.
2. A. Taflove and S. Hagness, *Computational Electrodynamics: The Finite Difference Time Domain Method*, 2nd edn., Artech House, 2000.
3. H. Mosallaei and Y. Rahmat-Samii, "Periodic bandgap and effective dielectric materials in electromagnetics: characterization and applications in nanocavities and waveguides," *IEEE Trans. Antennas Propagat.*, **vol. 51, no. 3,** 549–63, 2003.
4. F.-R. Yang, K.-P. Ma, Y. Qian, and T. Itoh, "A novel TEM waveguide using uniplanar compact photonic-bandgap (UC-PBG) structure," *IEEE Trans. Microwave Theory Tech.*, **vol. 47, no. 11,** 2092–8, 1999.

5. R. Cocciolo, F. R. Yang, K. P. Ma, and T. Itoh, "Aperture coupled patch antenna on UC-PBG substrate," *IEEE Trans. Microwave Theory Tech.*, **vol. 47**, 2123–30, 1999.
6. A. Aminian, F. Yang, and Y. Rahmat-Samii, "In-phase reflection and EM wave suppression characteristics of electromagnetic band gap ground planes," *2003 IEEE APS Int. Symp. Dig.*, vol. 4, pp. 430–3, June 2003.
7. F. Yang and Y. Rahmat-Samii, "Polarization dependent electromagnetic band-gap surfaces: characterization, designs, and applications," *2003 IEEE APS Int. Symp. Dig.*, vol. 3, pp. 339–42, June 2003.
8. F. Yang and Y. Rahmat-Samii, "Polarization dependent electromagnetic band gap (PDEBG) structures: designs and applications," *Microwave Optical Tech. Lett.*, **vol. 41**, no. 6, 439–44, 2004.
9. S. Maci, G. B. Gentili, P. Piazzesi, and C. Salvador, "Dual-band slot-loaded patch antenna," *IEE Proceedings on Microwaves Antennas & Propagation*, vol. 142, no. 3, pp. 225–32, June 1995.
10. X.-X. Zhang and F. Yang, "The study of slit cut on the microstrip antenna and its applications," *Microwave Optical and Tech. Lett.*, **vol. 18**, no. 4, 297–300, 1998.
11. F. Yang and X.-X. Zhang, "Slitted small microstrip antenna," *IEEE APS Int. Symp. Dig.*, pp. 1236–9, June 1998.
12. B. A. Munk, *Frequency Selective Surfaces: Theory and Design*, John Wiley & Sons, Inc., 2000.
13. F. Yang and Y. Rahmat-Samii, "A low profile single dipole antenna radiating circularly polarized waves," *IEEE Trans. Antennas Propagat.*, **vol. 53**, **no. 9**, 3083–6, 2005.
14. C. Balanis, *Advanced Engineering Electromagnetics*, Wiley, 1989.
15. J. McVay and N. Engheta, "High impedance metamaterial surfaces using Hilbert-curve inclusions," *IEEE Microw. Wireless Co. Lett.*, **vol. 14**, **no. 3**, 130–2, 2004.
16. C. R. Simovski, P. Maagt, and I. Melchakova, "High-impedance surfaces having stable resonance with respect to polarization and incidence angle," *IEEE Trans. Antennas Propagat.*, **vol. 53, no. 3**, 908–14, 2005.
17. M. A. Hiranandani, A. B. Yakovlev, and A. A. Kishk, "Artificial magnetic conductors realised by frequency-selective surfaces on a grounded dielectric slab for antenna applications," *IEE Proc. Microw. Antennas Propag.*, **vol. 153, no. 5**, 487–93, 2006.
18. L. Yang, M. Fan, and Z. Feng, "A spiral Electromagnetic Bandgap (EBG) structure and its application in microstrip antenna arrays," *2005 APMC Proceedings*, vol. 3, December 2005.
19. F. Yang, J. Chen, Q. Rui, and A. Elsherbeni, "A simple and efficient FDTD/PBC algorithm for periodic structure analysis," *Radio Sci.*, **vol. 42, no. 4**, RS4004, 2007.
20. *Ansoft Designer*, version 2.0, Ansoft Corporation, 2004.
21. Y. Kim, F. Yang, and A. Elsherbeni, "Compact artificial magnetic conductor designs using planar square spiral geometry," *Progress In Electromagnetics Research*, PIER **77**, 43–54, 2007.
22. D. F. Sievenpiper, *High-Impedance Electromagnetic Surfaces*, Ph.D. dissertation at University of California, Los Angeles, 1999.
23. J. Kennedy and R. Eberhart, "Particle swarm optimization," *Proc. 1995 Int. Conf. Neural Networks*, vol. IV, pp. 1942–8.
24. J. Robinson and Y. Rahmat-Samii, "Particle swarm optimization in electromagnetics," *IEEE Trans. Antennas Propagat.*, **vol. 52, no. 2**, 397–407, 2004.
25. Y. Rahmat-Samii, D. Gies, and J. Robinson, "Particle swarm optimization (PSO): a novel paradigm for antenna designs," *The Radio Science Bulletin*, **vol. 305**, 14–22, September 2003.

26. Y. Rahmat-Samii and E. Michielssen eds., *Electromagnetic Optimization by Genetic Algorithms*, John Wiley & Sons Inc., 1999.
27. N. Jin and Y. Rahmat-Samii, "Advances in particle swarm optimization for antenna designs: real-number, binary, single-objective and multiobjective implementations," *IEEE Trans. Antennas Propagat.*, **vol. 55, no. 3**, 556–67, 2007.
28. A. Tavallaee and Y. Rahmat-Samii, "A novel strategy for broadband and miniaturized EBG designs: hybrid MTL theory and PSO algorithm," *IEEE APS Int. Symp. Dig.*, pp. 161–4, June 2007.
29. N. Jin and Y. Rahmat-Samii, "Parallel particle swarm optimization and finite difference time-domain (PSO/FDTD) algorithm for multiband and wide-band patch antenna designs," *IEEE Trans. Antennas Propagat.*, **vol. 53, no. 11**, 3459–68, 2005.
30. N. Jin and Y. Rahmat-Samii, "Particle swarm optimization of miniaturized quadrature reflection phase structure for low-profile antenna applications," *IEEE APS Int. Symp. Dig.*, vol. 2B, pp. 255–8, July 2005.
31. A. Aminian, F. Yang, and Y. Rahmat-Samii, "Bandwidth determination for soft and hard ground planes by spectral FDTD: a unified approach in visible and surface wave regions," *IEEE Trans. Antennas Propagat.*, **vol. 53, no. 1**, 18–28, 2005.
32. J. McVay, N. Engheta, and A. Hoorfar, "Chapter 14: Space-filling curve high-impedance ground planes," in *Metamaterials: Physics and Engineering Explorations*, edited by N. Engheta and R. Ziolkowski, John Wiley & Sons Inc., 2006.
33. J. Zhu, A. Hoorfar, and N. Engheta, "Peano antennas," *IEEE Antennas Wireless Propag. Lett.*, **vol. 3**, 71–4, 2004.
34. J. McVay, A. Hoorfar, and N. Engheta, "Radiation characteristics of microstrip dipole antennas over a high-impedance metamaterial surface made of Hilbert inclusions," *Dig. 2003 IEEE MTT Int. Microwave Symp.*, pp. 587–90.
35. D. J. Kern, D. H. Werner, A. Monorchio, L. Lanuzza, and M. J. Wilhelm, "The design synthesis of multiband artificial magnetic conductors using high impedance frequency selective surfaces," *IEEE Trans. Antennas Propag*, **vol. 53, no. 1, part 1**, 8–17, 2005.
36. D. Sievenpiper, J. Schaffner, B. Loo, G. Tangonan, R. Harold, J. Pikulski, and R. Garcia, "Electronic beam steering using a varactor-tuned impedance surface," *IEEE APS Int. Symp. Dig.*, vol. 1, pp. 174–7, July 2001.
37. T. Liang, L. Li, J. A. Bossard, D. H. Werner, T. S. Mayer, "Reconfigurable ultra-thin EBG absorbers using conducting polymers," *IEEE APS Int. Symp. Dis.*, vol. 2B, pp. 204–7, 3–8 July 2005.
38. D. Sievenpiper, "Chapter 11: Review of theory, fabrication, and applications of high impedance ground planes," in *Metamaterials: Physics and Engineering Explorations*, edited by N. Engheta and R. Ziolkowski, John Wiley & Sons Inc., 2006.
39. D. F. Sievenpiper, J. H. Schaffner, H. J. Song, R. Y. Loo, and G. Tangonan, "Two-dimensional beam steering using an electrically tunable impedance surface," *IEEE Trans. Antennas Propagat.*, **vol. 51, no. 10**, 2713–22, 2003.

5 Patch antennas with EBG structures

Electromagnetic band gap structures have been characterized and designed in previous chapters. We now shift our focus to EBG applications in antenna engineering. In this chapter, the EBG structures are integrated into microstrip patch antenna designs and their surface wave band gap property helps to increase the antenna gain, minimize the back lobe, and reduce mutual coupling in array elements. Some applications of EBG patch antenna designs in high precision GPS receivers, wearable electronics, and phased array systems are highlighted at the end of the chapter.

5.1 Patch antennas on high permittivity substrate

Microstrip patch antennas are widely used in wireless communications due to the advantages of low profile, light weight, and low cost [1–2]. In principle, the microstrip patch antenna is a resonant type antenna, where the antenna size is determined by the operating wavelength and the bandwidth is determined by the Q factor of the resonance. An important research topic in microstrip antenna designs is to broaden the inherent narrow bandwidth of microstrip antennas. Parasitic patches are used to form a multi-resonant circuit so that the operating bandwidth is improved. In [3], the parasitic patches are located on the same layer with the main patch. In [4], a multi-layer microstrip antenna is investigated with parasitic patches stacked on the top of the main patch. The multi-resonant behavior can also be realized by incorporating slots into the metal patch. Several single-layer single-patch microstrip antennas have been reported, such as the U-slot microstrip antenna [5] and the E-shaped patch antenna [6].

Another important topic in microstrip antenna designs is to miniaturize the patch antenna size. The conventional half-wavelength size is relatively large in modern portable communication devices. Various approaches have been proposed, such as using shorting pins [7], cutting slots [8], and designing meandering microstrip lines [9]. Increasing the dielectric constant of the substrate is also a simple and effective way in reducing the antenna size [10].

Applications of microstrip antennas on high dielectric constant substrate are of growing interest due to their compact size and conformability with monolithic microwave integrated circuits (MMIC). However, there are several drawbacks with the use of high dielectric constant substrate, namely, narrow bandwidth, low radiation efficiency, and poor radiation patterns, which result from strong surface waves excited in the substrate.

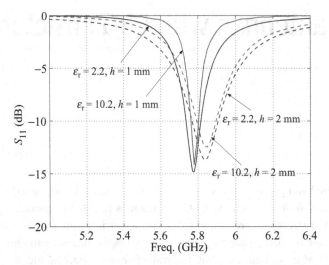

Fig. 5.1 Return loss comparison of patch antennas with different dielectric constants and substrate thicknesses.

The narrow bandwidth can be expanded by increasing the substrate thickness, which, however, will launch stronger surface waves. As a result, the radiation efficiency and patterns of the antenna are further degraded.

To quantify this phenomenon, a comparative study of patch antennas on substrates with different dielectric constants and different thicknesses is performed in this section [11]. Four cases are studied numerically using the finite difference time domain technique [12]:

$$(1)\ \varepsilon_r = 2.2, \quad h = 1\,\text{mm} \tag{5.1}$$

$$(2)\ \varepsilon_r = 2.2, \quad h = 2\,\text{mm} \tag{5.2}$$

$$(3)\ \varepsilon_r = 10.2, \quad h = 1\,\text{mm} \tag{5.3}$$

$$(4)\ \varepsilon_r = 10.2, \quad h = 2\,\text{mm} \tag{5.4}$$

Figure 5.1 shows the simulated S_{11} of these four structures. By tuning the patch size and the feeding probe location, all the antennas match well to 50 Ω around 5.8 GHz. The patch sizes are (16 mm, 9 mm), (15.5 mm, 12 mm), (7.5 mm, 5 mm), and (7 mm, 4 mm), respectively. It is noticed that the patch sizes on high dielectric constant substrate are remarkably smaller than those on low dielectric constant substrate, which is the main advantage of using high dielectric constant substrate. However, the antenna bandwidth ($S_{11} < 10$ dB) on 1 mm thick substrate is decreased from 1.38% to 0.61% when the dielectric constant is increased from 2.2 to 10.2. A similar phenomenon occurs for the 2 mm thickness cases: the bandwidth is decreased from 2.40% to 1.71%. In these results it is assumed that the dielectric substrate is lossless.

For the same dielectric constant substrates, the antenna bandwidth is enhanced when the thickness is doubled. For example, the antenna bandwidth on the high dielectric constant substrate is increased from 0.61% to 1.71% when the substrate thickness is

5.1 Patch antennas on high permittivity substrate

Table 5.1 Comparisons of patch antennas with different dielectric constants and different substrate thicknesses.

Case	ε_r	h (mm)	Patch size (mm)	Bandwidth (%)	Directivity (dB)
1	2.2	1	(16, 9)	1.38%	8.01
2	2.2	2	(15.5, 12)	2.40%	7.86
3	10.2	1	(7.5, 5)	0.61%	6.37
4	10.2	2	(7, 4)	1.71%	5.72

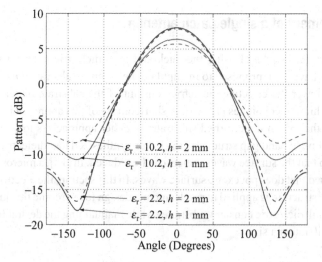

Fig. 5.2 Radiation pattern comparison of patch antennas with different dielectric constants and substrate thicknesses.

increased from 1 mm to 2 mm. It's worthwhile to point out that the bandwidth of case (4) is even larger than that of case (1), which means the bandwidth of microstrip antennas on high permittivity substrate can be recovered by increasing the substrate thickness. Some key values are listed in Table 5.1.

Figure 5.2 compares the H plane radiation patterns of these four antennas. Note that a finite ground plane of $1\lambda \times 1\lambda$ size is used in the simulations, where λ is the free space wavelength at 5.8 GHz. The antennas on the high dielectric constant substrates exhibit lower directivities and higher back radiation lobes than those on the low dielectric substrates. For antennas on the same dielectric constant substrate, when the thickness increases, the antenna directivity decreases, especially for those on high dielectric constant substrates. Similar observations are also found in the E plane patterns.

These phenomena can be explained from the excitation of surface waves in the substrate. When a high dielectric constant and thick substrate is used, strong surface waves are excited. This causes reduction of the radiation efficiency and the directivity. In addition, when the surface waves diffract at the edges of the ground plane, the back radiation is typically increased.

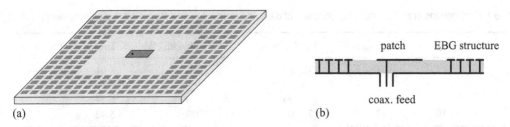

Fig. 5.3 Patch antenna surrounded by a mushroom-like EBG structure: (a) geometry and (b) cross section.

5.2 Gain enhancement of a single patch antenna

To overcome the drawbacks of using the thick and high dielectric constant substrate, several methods have been proposed to manipulate the antenna substrate. One approach suggested is to lower the effective dielectric constant of the substrate under the patch using micromachining techniques [13–14]. A shortcoming of this approach is the larger patch size than that on an unperturbed substrate. Another approach is to surround the patch with a complete band gap structure [15–16] or synthesized low dielectric constant substrate [17] so that the surface waves impact can be reduced. In [18], a microstrip patch design is proposed that does not excite surface waves. In this section, the electromagnetic band gap (EBG) structure is applied in patch antenna design to overcome the undesirable features of the high dielectric constant substrates while maintaining the desirable features of utilizing small antenna size.

5.2.1 Patch antenna surrounded by EBG structures

Figure 5.3 sketches the geometry of a microstrip patch antenna surrounded by a mushroom-like electromagnetic band gap (EBG) structure. The EBG is designed so that its surface wave band gap covers the antenna resonant frequency. As a result, the surface waves excited by the patch antenna are inhibited from propagation by the EBG structure. To effectively suppress the surface waves, four rows of EBG cells are used in the design. It is worthwhile to point out that the EBG cell is very compact because of the high dielectric constant and the thick substrate employed. Therefore the ground plane size can remain small, such as $1\lambda \times 1\lambda$.

For comparison purposes, another patch antenna designed on a step-like substrate is investigated, as shown in Fig. 5.4. The idea is to use a thick substrate under the patch which helps to keep the compact size and antenna bandwidth and use a thin substrate around the patch which could reduce the surface waves. This substrate geometry is like a stair step. The distance between the patch and the step needs to be carefully chosen. If the distance is too small, the resonance feature of the patch will change and the bandwidth will decrease. If the distance is too large, it cannot reduce the surface waves effectively.

To validate the above design concepts, four antennas are fabricated on RT/duroid 6010 ($\varepsilon_r = 10.2$) substrate with a finite ground plane of 52 mm × 52 mm ($1\lambda \times 1\lambda$). Two of them are normal patch antennas built on 1.27 mm and 2.54 mm thick substrates

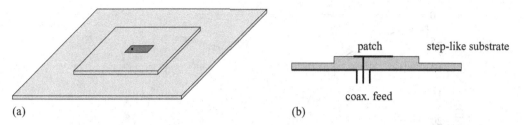

Fig. 5.4 Patch antenna on a step-like substrate: (a) geometry and (b) cross section.

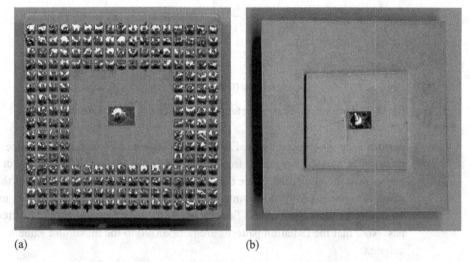

Fig. 5.5 Photos of two patch antenna designs with enhanced performance: (a) antenna surrounded by an EBG structure and (b) antenna on a step-like substrate.

as references. The step-like structure stacks two 1.27 mm thick substrates under the patch and the distance from the patch edge to the step is 10 mm. The EBG structure is built on 2.54 mm thick substrate and the EBG patch size is 2.5 mm × 2.5 mm with 0.5 mm separation. Figure 5.5 shows the photos of the EBG and step-like antenna prototypes.

Figure 5.6 compares the measured S_{11} results of these four antennas. All the four patches are tuned to resonate at the same frequency 5.8 GHz. It is noticed that the patch on the thin substrate has the narrowest bandwidth of only 1% while the other three have similar bandwidths of about 3–4%. Thus, the thickness of the substrate under the patch is the main factor determining the impedance bandwidth of the antenna. The step substrate and the EBG structure, which are located away from the patch antenna, have less effect on the antenna bandwidth.

Figure 5.7 presents the radiation patterns of these antennas, both in the E plane and H plane. The antenna on the thick substrate has the lowest front radiation while its back radiation is the largest. When the substrate thickness is reduced, the surface waves become weaker and the radiation pattern improves. The step-like structure exhibits similar radiation performance as the antenna on the thin substrate. The best radiation

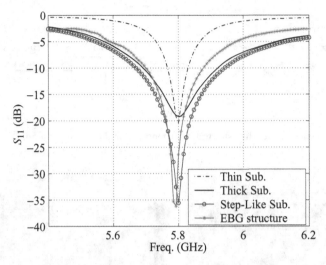

Fig. 5.6 Comparison of measured return loss of four patch antenna structures.

performance is achieved by the EBG antenna structure. Because of the successful suppression of surface waves, its front radiation is the highest, which is about 3.24 dB higher than the thick case. Since the surface wave diffraction at the edges of the ground plane is suppressed, the EBG antenna has a very low back lobe, which is more than 10 dB lower than other cases. Table 5.2 lists the experimental results of these antennas. Note that the radiation patterns are normalized to the maximum value of the EBG antenna.

It is also interesting to notice that in the E plane the beamwidth of the EBG case is much narrower than the other three cases whereas in the H plane it is similar to other designs. The reason is that the surface waves are mainly propagating along the E plane. Once the EBG structure stops the surface wave propagation, the beam becomes much narrower in the E plane.

From above comparisons it is clear that the EBG structure improves the radiation performances of the patch antenna while maintaining its compact size and adequate bandwidth. Since the cavity structure is often used to improve radiation patterns, a cavity back patch antenna is also simulated and compared to the EBG antenna [19]. The FDTD computed radiation patterns are shown in Fig. 5.8. Although cavity structure exhibits some improvements over the conventional patch antenna, its performance is not as good as the EBG structure.

5.2.2 Circularly polarized patch antenna design

The aforementioned patch antenna is linearly polarized. The EBG structure is also used to improve the performance of circularly polarized patch antennas [20–21]. As shown in Fig. 5.9, a probe fed square patch antenna with a size of 6.75 mm × 6.75 mm is designed on a 1.575 mm thick substrate with a dielectric constant of 4.7. A small perturbation is added in the corner and the probe location is selected to properly excite both the TM_{01}

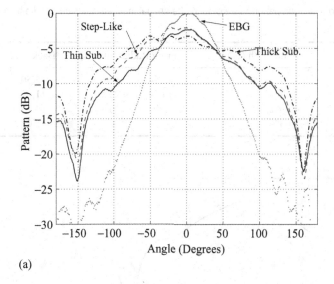

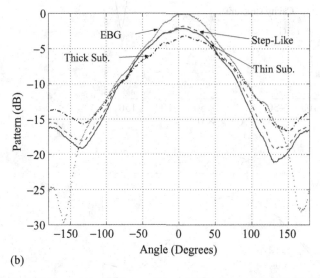

Fig. 5.7 Measured radiation patterns of different patch antennas: (a) E plane pattern and (b) H plane pattern.

and TM_{10} modes for circular polarization. The patch is surrounded by two rows of EBG structures. Dimensions of the EBG structure are designed such that the band gap can accommodate the operating frequency band of the antenna. The period of the cell is 3.25 mm with a gap of 0.25 mm between the square patches. The diameter of the shorting pin is 0.5 mm. The total ground plane size is 26.25 mm × 26.25 mm.

The return loss, axial ratio, and radiation pattern characteristics of the EBG patch antenna are presented in [20]. A significant improvement in all the characteristics is evident from the comparison with a reference patch antenna without EBG. For example,

Table 5.2 Measured antenna performance of four different patch antenna designs on the high dielectric constant substrate.

Antennas	Bandwidth (%)	Front Radiation (dB)	Back Radiation (dB)
Thin	1.07%	−2.34	−15.50
Thick	3.93%	−3.24	−11.92
Step-Like	4.69%	−2.03	−14.19
EBG	3.03%	0	−25.0

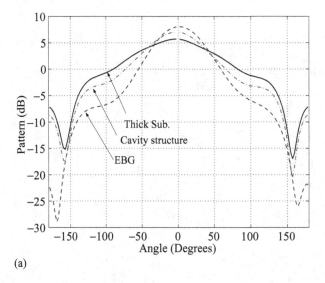

(a)

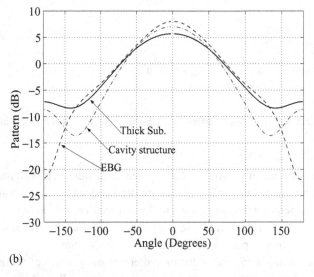

(b)

Fig. 5.8 Comparisons of simulated radiation patterns of patch antennas surrounded by the EBG structure and cavity structure: (a) E plane pattern and (b) H plane pattern.

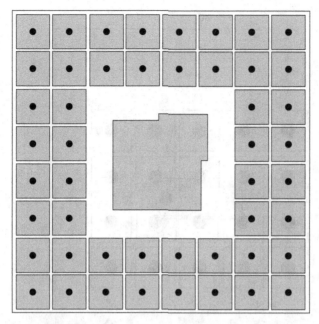

Fig. 5.9 Geometry of a modified square patch antenna surrounded by EBG cells for circular polarization (from [20], © IEEE 2001).

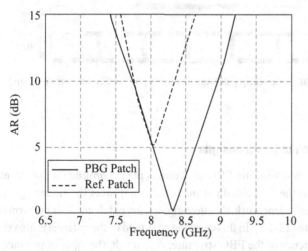

Fig. 5.10 Axial ratio of the EBG patch antenna compared with a reference patch antenna (from [20], © IEEE 2001).

Fig. 5.10 shows the axial ratio result. When AR $\leq$ 6 dB is used as a criterion, the corresponding bandwidth is increased from 1.5% to 9.6% [20]. Also an improvement in the front to back ratio of the radiation patterns has been observed due to the reduction of surface wave diffractions at the edges of the finite ground plane.

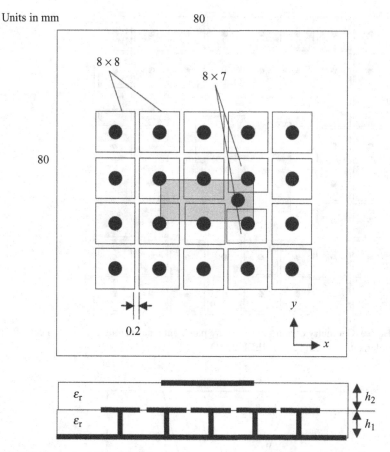

Fig. 5.11 Geometry of a microstrip patch antenna on top of an EBG substrate (from [23], © IET 2006).

5.2.3 Various EBG patch antenna designs

In the previous discussions, the EBG structure is placed *around* the patch antenna. The EBG mainly affects the far field of the antenna and improves the antenna gain and radiation patterns. Some recent work also investigates the patch antenna performance *on top of* an EBG substrate [22–23]. In this case, the near field of the antenna is also significantly altered by the existence of the EBG structure. As a result, the input impedance and bandwidth also change. For example, Fig. 5.11 shows a two-layer EBG patch antenna design [23]. The patch length is 14 mm and width is 6 mm. The thickness for each substrate layer is 1.59 mm and the dielectric constant is 2.5. The dimensions of the EBG structures are labeled in the figure.

A parametric study was conducted for this antenna, including the feed location, patch width, ground plane size, and EBG spacing. The results show that a microstrip patch antenna over an EBG ground plane has a significantly improved performance in its impedance bandwidth, gain, and cross polarization. Figure 5.12 presents the

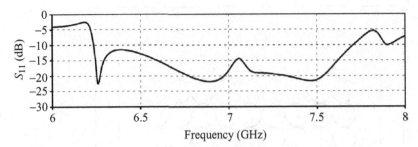

Fig. 5.12 S_{11} of the EBG patch antenna shown in Fig. 5.11 with EBG patch spacing of 0.3 mm (from [23], © IET 2006).

Fig. 5.13 Photograph of a prototype patch antenna surrounded by three rows of UC-EBG cells (from [15], © IEEE 1999).

simulated antenna return loss, which has a broader bandwidth than the conventional patch antenna [23]. In order to confirm the simulated results, several antennas were fabricated, and the reflection coefficient S_{11} and the E- and H-plane radiation patterns were measured. A 13.66% bandwidth and 9.77 dB gain were obtained for the fabricated antenna [23].

Besides the mushroom-like EBG structure, other EBG structures are also used to improve the performance of patch antennas [15–16, 24]. For example, Fig. 5.13 shows a photograph of an aperture coupled patch antenna surrounded by a uni-planar compact EBG structure [15]. The patch is etched on a standard 50-mil-thick substrate with a

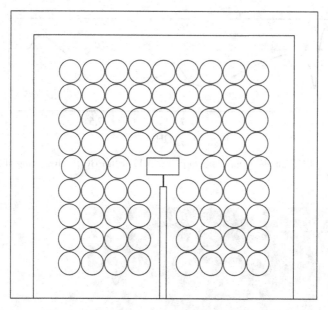

Fig. 5.14 Patch antenna on an EBG substrate formed by four holes in a square lattice (from [16], © IEEE 1999).

dielectric constant of 10.2. The patch length is 210 mil and width is 100 mil, resulting in a resonant frequency at 12 GHz. The patch antenna is surrounded by three rows of UC-EBG cells. A 50-mil gap is left between the patch and UC-EBG metal pattern. Measured data show achievement of a more focused beam radiated in the broadside direction with over 3-dB gain enhancement.

Figure 5.14 shows another EBG patch antenna design, where the EBG structure is a square lattice of air columns embedded in a dielectric medium [16]. The dielectric substrate has a dielectric constant of 10.2 and thickness of 10 mm. The lattice period is 38 mm and the ratio of column radius to periodicity is 0.48. The patch antenna is rectangular with 52-mm width and 25.96-mm length, and it resonates around 2.2 GHz. The transmission line, designed to feed the patch antenna, is formed by a 50 Ω line together with a quarter-wavelength transformer for impedance matching. Comparisons between a conventional patch antenna and an EBG patch antenna show that the reduction in the surface wave level is remarkable. Therefore, the ripples in the radiation pattern due to surface waves have almost disappeared. The back radiation is also considerably reduced from −12 dB to −24 dB.

5.3 Mutual coupling reduction of a patch array

After discussing EBG applications in single microstrip patch antennas, we now present how EBG can help to improve the performance of microstrip antenna arrays. As an important parameter in array design, the mutual coupling between array elements is

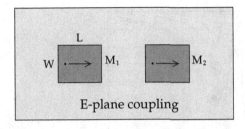

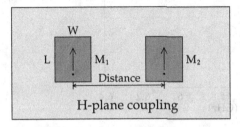

Fig. 5.15 The E-plane and H-plane coupled probe fed microstrip antenna arrays (from [25], © IEEE 2003).

investigated both numerically and experimentally. Strong mutual coupling could reduce the array efficiency and cause the scan blindness in phased array systems. Therefore, the electromagnetic band gap (EBG) structures are used to reduce the coupling between array elements.

5.3.1 Mutual coupling between patch antennas on high dielectric constant substrate

Let's start with the mutual coupling characterization of conventional microstrip patch antennas on substrates with different thicknesses and permittivities. The same four cases as (5.1) to (5.4) are analyzed here and the antenna dimensions are the same as listed in Table 5.1. Both the E-plane and H-plane array structures are investigated. In the E-plane coupled case, the patch antennas are aligned parallel to the direction of the electric current on the patch, as shown in Fig. 5.15. In the H-plane coupled case, the patch antennas are aligned perpendicular to the electric current direction. The FDTD method is used to calculate the mutual coupling between the patch antennas [25].

The mutual coupling of the E-plane coupled microstrip antennas are depicted in Fig. 5.16. All the antennas are tuned to resonate around 5.8 GHz. The proper distance between the antenna elements guarantees that the return loss is very similar to Fig. 5.1. Figure 5.16a presents the mutual coupling of the microstrip antennas with a 0.50λ antenna separation, where λ is the free space wavelength at the resonant frequency 5.8 GHz. Note that the distance is measured from probe to probe, not between two adjacent edges of the patch antennas. It is observed that the antennas on the thin and low dielectric constant substrate have the lowest mutual coupling level while the antennas on the thick and high dielectric constant substrate show strongest mutual coupling. The mutual coupling becomes stronger when the dielectric constant is increased or the substrate thickness is increased. The reason is that the microstrip antenna on a high permittivity

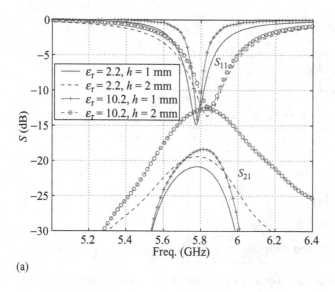

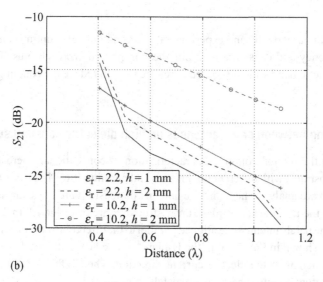

Fig. 5.16 Comparisons of the E-plane coupled microstrip antennas on different dielectric constants and different thickness substrates: (a) S_{11} and S_{21} vs. frequency (distance = 0.5 λ), (b) S_{21} vs. distance (f = 5.8 GHz) (from [25], © IEEE 2003).

and thick substrate launches the most severe surface waves. Figure 5.16b plots how the mutual couplings at the resonant frequency vary with the patch distance. The mutual couplings of all cases decrease while the antenna distance increases.

The H-plane coupled microstrip antenna results are depicted in Fig. 5.17. In Fig. 5.17a, the distance between patch antennas is fixed at 0.50 λ, and the mutual coupling versus frequency is plotted. Figure 5.17b presents the mutual coupling variation with antenna distance at the resonant frequency 5.8 GHz. In contrast to the E-plane coupled

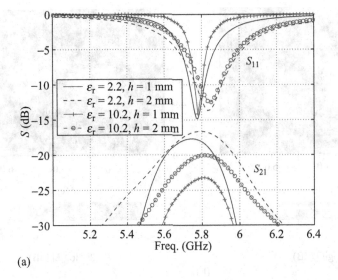

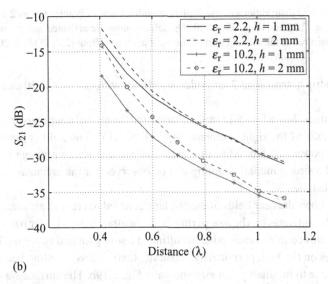

Fig. 5.17 Comparisons of the H-plane coupled microstrip antennas on different dielectric constants and different thickness substrates: (a) S_{11} and S_{21} vs. frequency (distance = 0.5 λ), (b) S_{21} vs. distance ($f = 5.8$ GHz) (from [25], © IEEE 2003).

results, the strongest mutual coupling happens when the antenna array is mounted on a low dielectric constant and thick substrate, and the weakest mutual coupling happens for the case with a high dielectric constant and thin substrate. It is interesting to notice that increasing the substrate thickness still increases the mutual coupling, but increasing the permittivity decreases the mutual coupling.

To understand the different coupling mechanisms of the E-plane and H-plane coupled microstrip antenna arrays, the near field distributions of different coupling situations are calculated and graphically presented. Figure 5.18 plots the near fields of the E-plane

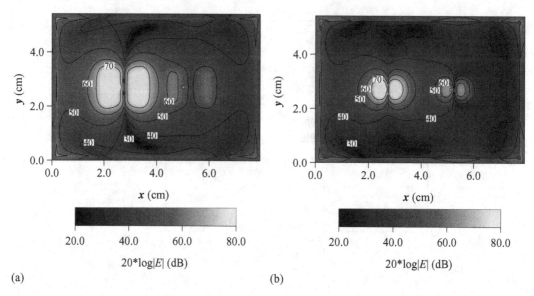

Fig. 5.18 The near field distributions of the E-plane coupled microstrip antennas on 2 mm thick substrates: (a) $\varepsilon_r = 2.20$ and (b) $\varepsilon_r = 10.20$. The left antennas are activated and the mutual couplings are measured at the feeding port of the right antennas (from [25], © IEEE 2003).

coupled microstrip antennas on 2 mm thick substrates with (a) $\varepsilon_r = 2.20$ and (b) $\varepsilon_r = 10.20$.

The left antennas (M_1 in Fig. 5.15) are activated and the mutual couplings are measured at the feeding port of the right antennas (M_2 in Fig. 5.15). The field is normalized to 1 W delivered power and plotted in dB scale. The surface waves propagate along the x direction and a strong mutual coupling can be observed for the antennas on the high permittivity substrate.

Figure 5.19 shows the near fields of the H-plane coupled microstrip antennas. As seen in Fig. 5.19a, the antennas on the low permittivity substrate have a larger patch size and their fringing fields couple to each other, resulting in a strong mutual coupling. However, for the antennas on the high permittivity substrate, there is less coupling between their fringing fields due to its small patch size shown in Fig. 5.19b. The surface waves which contribute to the strong mutual coupling of the E-plane coupled case have less effect now because their propagation direction is perpendicular to the array alignments.

It can be concluded from the above results that the mutual coupling of microstrip antennas is determined by both the directional surface waves and antenna size. In particular, the surface waves have a strong effect in the E-plane mutual coupling of the microstrip antenna array.

5.3.2 Mutual coupling reduction by the EBG structure

From the preceding investigations, we notice that the E-plane coupled microstrip antenna array on a thick and high permittivity substrate suffers from strong mutual coupling

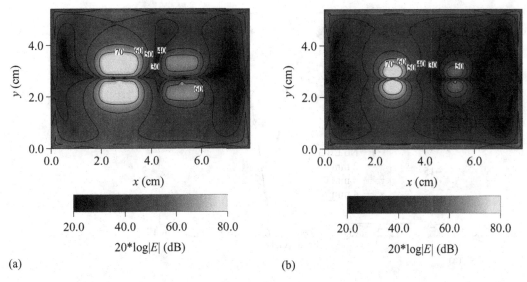

Fig. 5.19 The near field distributions of the H-plane coupled microstrip antennas on 2 mm thick substrates: (a) $\varepsilon_r = 2.20$ and (b) $\varepsilon_r = 10.20$. The left antennas are activated and the mutual couplings are measured at the feeding port of the right antennas (from [25], © IEEE 2003).

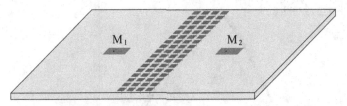

Fig. 5.20 Microstrip antennas separated by the mushroom-like EBG structure for a low mutual coupling (from [25], © IEEE 2003).

because of the pronounced surface waves. Since the EBG structure has already demonstrated its capability to suppress surface waves, it is inserted between the antennas to reduce the mutual coupling, as shown in Fig. 5.20. Note that four rows of EBG cells are used here to obtain a satisfactory result.

FDTD technique is used to simulate the E-plane coupled microstrip antennas on a dielectric substrate with $h = 2$ mm and $\varepsilon_r = 10.2$. The antenna's size is 7 mm × 4 mm and the distance between the antennas is 38.8 mm (0.75 λ). The mushroom-like EBG structure is inserted between the antennas to reduce the mutual coupling. Three different EBG cases are analyzed and their patch sizes are 2 mm, 3 mm, and 4 mm, respectively. The gap between mushroom-like patches is kept at 0.5 mm for all three cases.

Figure 5.21a shows the return loss of three EBG cases as well as the antennas without the EBG structure. It is observed that all the antennas resonate around 5.8 GHz. Although the existence of the EBG structure has some effects on the input matches of the antennas, all the antennas still have return losses better than -10 dB.

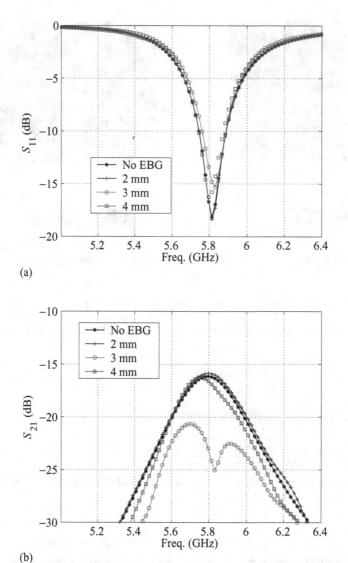

Fig. 5.21 FDTD simulated results of the E-plane coupled microstrip antennas separated by EBG structures with different mushroom-like patch sizes: (a) return loss and (b) mutual coupling (from [25], © IEEE 2003).

The mutual coupling results are shown in Fig. 5.21b. Without the EBG structure, the antennas show a strong mutual coupling of −16.15 dB. If the EBG structures are employed, the mutual coupling level changes. When the 2 mm EBG is used, its band gap is higher than the resonant frequency 5.8 GHz. Therefore, the mutual coupling is not reduced and a strong coupling of −15.85 dB is still noticed. For the 3 mm EBG case, the resonant frequency 5.8 GHz falls inside the EBG band gap so that the surface waves are suppressed. As a result, the mutual coupling is greatly reduced: only −25.03 dB at the resonant frequency. It is worthwhile to point out that the bandwidth of the

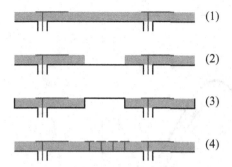

Fig. 5.22 Geometries of four different microstrip antenna array structures: (1) normal microstrip antennas, (2) substrate between antennas is removed, (3) cavity back microstrip antennas, and (4) microstrip antennas with the EBG structure in between (from [25], © IEEE 2003).

mushroom-like EBG structure is wider than the antenna bandwidth so that it can cover the entire operational band of the antenna. When the size of the mushroom-like patch is increased to 4 mm, its frequency band gap further decreases, and is now lower than the resonant frequency. Therefore, the mutual coupling is not improved and is still as strong as -16.27 dB.

It is instructive to compare the EBG structure with other structures also used to reduce the mutual coupling. Figure 5.22 plots four antenna structures to be compared:

(1) Normal microstrip antennas
(2) The substrate between antennas is removed
(3) Cavity back microstrip antennas
(4) Microstrip antennas with the EBG structure in between

For this comparison, the antenna size, substrate properties, and antenna distance are kept the same. In structure (2), a 13.5 mm width substrate is removed between the patch antennas. This width is chosen to be the same as the total width of four rows of the EBG patches. When the cavity structure is used, the distance between the adjacent PEC walls is also selected to be 13.5 mm.

Figure 5.23 displays the mutual coupling results of these four different structures. The normal microstrip antennas show the highest mutual coupling. The substrate removal case and the cavity back case have some improvements on reducing the mutual coupling. A 1.5 dB mutual coupling reduction is noticed for the former case and a 2 dB reduction is observed for the latter case. The lowest mutual coupling is obtained in the EBG case where an 8.8 dB reduction is achieved. This comparison demonstrates the unique capability of the EBG structure to reduce the mutual coupling.

To verify the conclusions drawn from the FDTD simulation, two pairs of microstrip antennas are fabricated on Roger RT/Duroid 6010 substrates. The permittivity of the substrate is 10.2, and the substrate thickness is 1.92 mm (75 mil). Figure 5.24 shows a photograph of the fabricated antennas with and without the EBG structure. The antenna's size is 6.8 mm × 5 mm, and the distance between the antennas' edges is 38.8 mm (0.75 λ). The antennas are fabricated on a ground plane of 100 mm × 50 mm. For the EBG structures, the mushroom-like patch size is 3 mm and the gap between the patches is 0.5 mm.

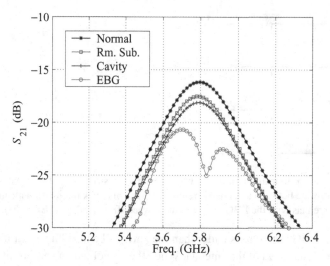

Fig. 5.23 Comparison of mutual coupling results of four different microstrip antenna structures (from [25], © IEEE 2003).

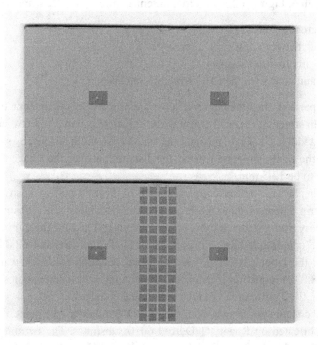

Fig. 5.24 A photograph of microstrip antenna arrays with and without the EBG structure. The substrate thickness is 1.92 mm and its dielectric constant is 10.2 (from [25], © IEEE 2003).

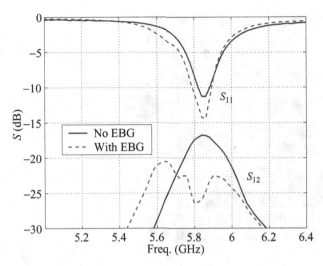

Fig. 5.25 Measured results of microstrip antennas with and without the EBG structure (from [25], © IEEE 2003).

Four columns of EBG cells are inserted between the antennas to reduce the mutual coupling.

The measured results are shown in Fig. 5.25. It is observed that both antennas resonate at 5.86 GHz with return loss better than -10 dB. For the antennas without the EBG structure, the mutual coupling at 5.86 GHz is -16.8 dB. In comparison, the mutual coupling of the antennas with the EBG structure is only -24.6 dB. An approximately 8 dB reduction of mutual coupling is achieved at the resonant frequency of 5.86 GHz. This result agrees well with the simulated result shown in Fig. 5.23. From this experimental demonstration, it can be concluded that the EBG structure can be utilized to reduce the antenna mutual coupling between array elements.

5.3.3 More design examples

The concept of using EBG structures to reduce the mutual coupling between patch antennas has been realized in different designs [26–31]. For example, a dumbbell EBG structure in patch antenna array designs is discussed in [26–27]. Fig. 5.26 shows a 2×2 microstrip patch antenna array built on a substrate of 2 mm thick with a dielectric constant of 4.6 [27]. A dumbbell EBG structure is designed to suppress the surface waves at the antenna resonant frequency of 5.6 GHz. The simulation and experiment results show that mutual coupling S_{21} between the elements is reduced by around 4 dB compared to the array without EBG. As a result, a 1.5 dB gain improvement and 1.7 dB side lobe level reduction are achieved in the EBG patch array design.

Another EBG patch array is designed using the uni-planar compact electromagnetic band gap (UC-EBG) structure. Probe fed circular patch antennas are built on a substrate with dielectric constant 10.2 and thickness 2.54 mm. Array prototypes with and without UC-EBG were manufactured and a photograph is shown in Fig. 5.27 [28]. The measured

Fig. 5.26 Photograph of a 2 × 2 rectangular patch antenna array with dumbbell EBG cells between array elements (from [27], © Wiley Inter-Science 2003).

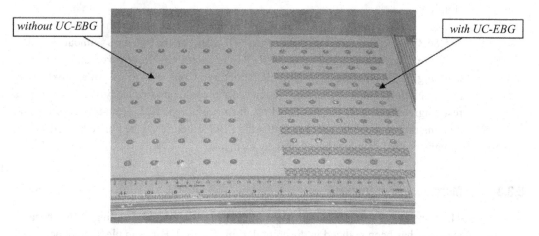

Fig. 5.27 Photograph of 7 × 5 circular patch antenna arrays with and without UC-EBG cells (from [28], © IEEE 2004).

data show that the mutual coupling is significantly higher in the E plane than in the H plane due to the radiation mechanism of the patch, which favors surface wave excitation in the E plane. After inserting two rows of UC-EBG cells between adjacent elements, the E-plane coupling is reduced from −12.6 dB to −18.7 dB at the resonant frequency 5.75 GHz. In addition, the coupling to the second adjacent elements is reduced from −24.5 dB to −42.8 dB.

5.4 EBG patch antenna applications

It is clear from previous sections that the EBG structures help to improve the performance of microstrip patch antennas, namely, increase the antenna gain and efficiency, suppress the back radiation, and reduce the mutual coupling. As a consequence, they have found broad applications in various communication systems [32–33]. In this section, several representative applications are reviewed, including the high precision GPS receiver, wearable electronics, and phased array system.

5.4.1 EBG patch antenna for high precision GPS applications

High precision GPS system allows surveyors to make static measurements with sub-centimeter accuracy. To obtain these accuracies, extra precautions are needed to shield the antenna from unwanted multi-path signals, such as the reflection from the ground. A traditional approach is to use choke rings, which provides excellent electrical performance. However, they are usually very large, heavy and costly. Recently, EBG structures have been used to replace the bulky rings while maintaining the good antenna performance [34–35]. Figure 5.28a shows a photograph of a GPS antenna on an EBG ground plane [33, 35]. The measured results demonstrate proper pattern shaping at both L_1 and L_2 bands, which allows good antenna operation above the horizon and good cross-polarization rejection below the horizon. Compared to a GPS antenna with choke rings, the EBG ground plane is about 7 times lighter, 5 times thinner, and 2 to 8 times less expensive [35].

Figure 5.28b shows another EBG patch antenna design for the GPS application [36], which is optimized for both a compact antenna size and good radiation performance. In this design, a patch antenna on a high permittivity substrate ($\varepsilon_r = 10.2$) is integrated with a fractal EBG surface and the total ground plane size is only 80 mm × 80 mm. It provides a good circular polarization within a reasonably wide frequency range that covers the L_2 band of the GPS standard. A gain improvement of 1.5 dB and a 60% wider axial-ratio bandwidth is achieved as compared to the same antenna without EBG.

5.4.2 EBG patch antenna for wearable electronics

In recent years, the development of wearable electronic systems has been rapid [37]. A variety of consumer electronics will be built into the garments of future clothing. A typical wearable system consists of CPU, LCD display, input devices, wireless communication links, positioning system, and batteries. Wearable antennas have received much interest due to the integration of personal communications into wearable electronics. They play a paramount role in the optimal design of the wearable system. Clearly in designing these antennas, the electromagnetic interaction among the antenna, wearable unit and the human operator is an important factor to be considered. EBG technology has been proposed as a potential design solution [38–39].

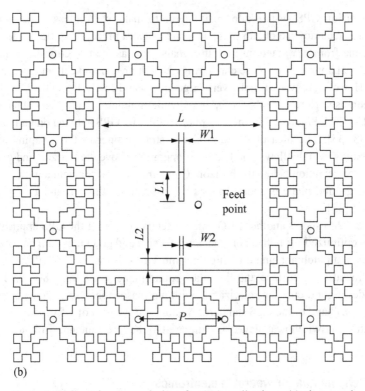

Fig. 5.28 EBG patch antenna designs for GPS applications: (a) photograph of a patch antenna on a metal dielectric EBG substrate (from [35], © IEEE 2002), (b) geometry of GPS patch antenna with the fractal EBG structure (from [36], © IEEE 2006).

A two-layer wearable EBG antenna was designed and fabricated, as shown in Fig. 5.29 [39]. The conventional felt fabric whose dielectric constant is 1.1 is used as antenna substrate. In the middle layer of the antenna 36 EBG cells are glued, and the main patch antenna is placed on top of the EBG substrate. The two layers are then sewn together and a coaxial SMA connector is used to feed the antenna. By introducing

(a)   (b)

Fig. 5.29 Photographs of a wearable EBG patch antenna: (a) patch antenna on the top layer and (b) EBG surface in the middle layer (from [39], © IEEE 2004).

the EBG type structures into a conventional textile antenna, the required antenna size is reduced by more than 30%. In addition, the input match bandwidth is notably increased.

5.4.3 EBG patch antennas in phased arrays for scan blindness elimination

Electronically scanned phased arrays find their use in many applications such as radar systems and constellations of low Earth orbit satellites. These applications require scanned multi-beam antennas with relatively wide bandwidth. The use of active phased array built by microstrip technology is an attractive solution. However, the need for the bandwidth and scan angle increases the undesirable effects caused by surface waves. In particular, scan blindness occurs when propagation constants of Floquet modes (space harmonics) coincide with those of surface waves supported by the array structure. The scan blindness limits the scan range and lowers the antenna efficiency.

Using EBG structures in the dielectric substrate gives a promising way to eradicate the problem, while at the same time improving antenna performance [40–42]. A detailed theoretical analysis and experiment verification are presented in [41]. Figure 5.30 compares the scanned performance of printed dipole arrays on EBG substrate and on a uniform dielectric substrate [41]. For the uniform substrate case a scan blindness occurs at 69.0° in the E plane. In contrast, there is no blind angle observed for the EBG case. Note that there is no blindness in the H plane or diagonal plane because of polarization mismatch. Similar analysis is performed for an array on a thick substrate. A thicker substrate provides a wider bandwidth but excites more severe surface waves. As shown in Fig. 5.30b, when the substrate thickness is doubled, the blind angle in the E-plane is reduced to 50°. After using the EBG substrate, the blind angle is completely removed. The same analysis is extended to microstrip patch array and similar observations are observed as well [42].

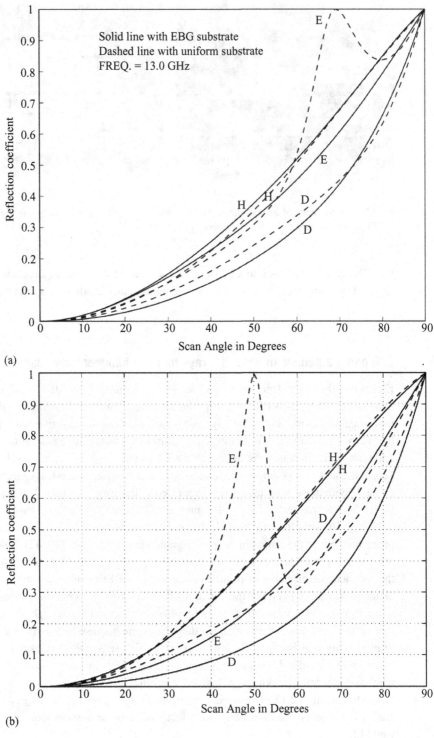

Fig. 5.30 Comparison of scanned performance of printed dipole arrays on EBG substrate and on uniform dielectric substrate: (a) substrate thickness is 2.39 mm and (b) substrate thickness is 4.77 mm (from [41], © IEEE 2004).

5.5 Projects

Project 5.1 Investigate the performance of a circular patch antenna on dielectric substrates with different thicknesses and dielectric constants. Compare your observations with the results obtained in Section 5.1.

Project 5.2 Four rows of EBG cells are implemented in Fig. 5.3 and Fig. 5.20. Carry out a parametric study on the number of EBG rows for patch antenna applications.

References

1. J. J. Bahl and P. Bhartia, *Microstrip Antennas*, Artech House, 1980.
2. P. Bhartia, Inder Bahl, R. Garg, and A. Ittipiboon, *Microstrip Antenna Design Handbook*, Artech House, 2000.
3. G. Kumar and K. C. Gupta, "Directly coupled multiple resonator wide-band microstrip antenna," *IEEE Trans. Antennas Propagat.*, **vol. AP-33**, 588–93, 1985.
4. D. M. Pozar, "Microstrip antenna coupled to a microstrip-line," *Electron. Lett.*, **vol. 21, no. 2**, 49–50, 1985.
5. T. Huynh and K. F. Lee, "Single-layer single-patch wideband microstrip antenna," *Electron. Lett.*, **vol. 31, no. 16**, 1310–12, 1995.
6. F. Yang, X.-X. Zhang, X. Ye, and Y. Rahmat-Samii, "Wideband E-shaped patch antennas for wireless communications," *IEEE Trans. Antennas Propagat.*, **vol. 49, no. 7**, 1094–100, 2001.
7. A. K. Shackelford, K.-F. Lee, and K. M. Luk, "Design of small-size wide-bandwidth microstrip-patch antennas," *IEEE Antennas and Propagat. Magazine*, **vol. 45, no. 1**, 75–83, 2003.
8. X.-X. Zhang and F. Yang, "The study of slit cut on the microstrip antenna and its applications," *Microwave Optical Tech. Lett.*, **vol. 18, no. 4**, 297–300, 1998.
9. S. Dey and R. Mittra, "Compact microstrip patch antenna", *Microwave Optical Tech. Lett.*, **vol. 12, no. 1**, 12–14, 1996.
10. T. K. Lo, C.-O. Ho, Y. Hwang, E. K. W. Lam, and B. Lee, "Miniature aperture coupled microstrip antenna of very high permittivity," *Electron. Lett.*, **vol. 33**, 9–10, 1997.
11. F. Yang, *EBG Structure and Reconfigurable Technique in Antenna Designs: Applications to Wireless Communications*, Ph.D. dissertation at University of California, Los Angeles, 2002.
12. M. A. Jensen, *Time-Domain Finite-Difference Methods in Electromagnetics: Application to Personal Communication*, Ph.D. dissertation at University of California, Los Angeles, 1994.
13. G. P. Gauthier, A. Courtay, and G. H. Rebeiz, "Microstrip antennas on synthesized low dielectric-constant substrate," *IEEE Trans. Antennas Propagat.*, **vol. 45**, 1310–14, 1997.
14. I. Papapolymerou, R. F. Frayton, and L. P. B. Katehi, "Micromachined patch antennas," *IEEE Trans. Antennas Propagat.*, **vol. 46**, 275–83, 1998.
15. R. Coccioli, F. R. Yang, K. P. Ma, and T. Itoh, "Aperture-coupled patch antenna on UC-PBG substrate," *IEEE Trans. Microwave Theory Tech.*, **vol. 47**, 2123–30, 1999.
16. R. Gonzalo, P. Maagt, and M. Sorolla, "Enhanced patch-antenna performance by suppressing surface waves using photonic-bandgap substrates," *IEEE Trans. Microwave Theory Tech.*, **vol. 47**, 2131–8, 1999.
17. J. S. Colburn and Y. Rahmat-Samii, "Patch antennas on externally perforated high dielectric constant substrates," *IEEE Trans. Antennas Propagat.*, **vol. 47**, 1785–94, 1999.

18. D. R. Jackson, J. T. Williams, A. K. Bhattacharyya, R. L. Smith, S. J. Buchheit, and S. A. Long, "Microstrip patch antenna designs that do not excite surface waves," *IEEE Trans. Antennas Propagat.*, **vol. 41**, 1026–37, 1993.
19. F. Yang and Y. Rahmat-Samii, "Step-Like structure and EBG Structure to improve the performance of patch antennas on high dielectric substrate," in *Proc. IEEE APS Dig.*, vol. 2, 2001, pp. 482–5.
20. M. Rahman and M. Stuchly, "Wide-band microstrip patch antenna with planar PBG structure," in *Proc. IEEE APS Dig.*, vol. 2, 2001, 486–9.
21. M. Rahman and M. A. Stuchly, "Circularly polarized patch antenna with periodic structure," *IEE Proc. Microwaves, Antennas Propagation*, **vol. 149, issue 3**, 141–6, 2002.
22. M. Y. Fan, R. Hu, Z. H. Feng, X. X. Zhang, and Q. Hao, "Advance in 2D-EBG research," *J. Infrared Millimeter Waves.*, **vol. 22, no. 2**, 2003.
23. D. Qu, L. Shafai, and A. Foroozesh, "Improving microstrip patch antenna performance using EBG substrates," *IEE Proc. Microwaves, Antennas Propagation*, **vol. 153, issue 6**, 558–63, 2006.
24. M. N. Mollah and N. C. Karmakar, "Planar PBG structures and their applications to antennas," *Proc. IEEE APS Dig.*, vol. 2, July 2001, pp. 494–7.
25. F. Yang and Y. Rahmat-Samii, "Microstrip antennas integrated with electromagnetic band-gap (EBG) structures: a low mutual coupling design for array applications," *IEEE Trans. Antennas Propagat.*, **vol. 51, no. 10**, 2936–46, 2003.
26. A. Yu and X.-X. Zhang, "A novel method to improve the performance of microstrip antenna arrays using a dumbbell EBG structure," *IEEE Antennas Wireless Propagat. Lett.*, **vol. 2**, 170–2, 2003.
27. N. Jin, A. Yu, and X.-X. Zhang, "An enhanced 2 × 2 antenna array based on a dumbbell EBG structure," *Microwave Optical Tech. Lett.*, **vol. 39, no. 5**, 395–9, 2003.
28. Z. Iluz, R. Shavit, and R. Bauer, "Microstrip antenna phased array with Electromagnetic bandgap substrate," *IEEE Trans. Antennas Propagat.*, **vol. 52, no. 6**, 1446–53, 2004.
29. L. Yang, Z. H. Feng, F. L. Chen, and M. Y. Fan, "A novel compact electromagnetic band-gap (EBG) structure and its application in microstrip antenna arrays," *IEEE MTT-S Int. Microwave Symp. Dig.*, pp. 1635–8, 2004.
30. Y. Yao, X. Wang, and Z. Feng, "A novel dual-band compact electromagnetic bandgap (EBG) structure and its application in multi-antennas," in *Proc. IEEE APS Dig.*, pp. 1943–6, 2006.
31. K. Buell, H. Mosallaei, and K. Sarabandi, "Metamaterial insulator enabled superdirective array," *IEEE Trans. Antennas Propagat.*, **vol. 55, no. 4**, 1074–85, 2007.
32. F. Yang and Y. Rahmat-Samii, "Applications of electromagnetic band-gap (EBG) structures in microwave antenna designs," *Proc. of 3rd International Conference on Microwave and Millimeter Wave Technology*, 528–31, 2002.
33. P. de Maagt, R. Gonzalo, Y. C. Vardaxoglou, and J.-M. Baracco, "Electromagnetic bandgap antennas and components for microwave and (sub)millimeter wave applications," *IEEE Trans. Antennas Propagat.*, **vol. 51, no. 10**, 2667–77, 2003.
34. R. Hurtado, W. Klimczak, W. E. McKinzie, and A. Humen, "Artificial magnetic conductor technology reduces weight and size for precision GPS antennas," *Navigational National Technical Meeting*, San Diego, CA, January 28–30, 2002.

35. W. E. McKinzie III, R. B. Hurtado, B. K. Klimczak, and J. D. Dutton, "Mitigation of multipath through the use of an artificial magnetic conductor for precision GPS surveying antennas," *Proc. IEEE APS Dig.*, vol. 4, pp. 640–3, 2002.
36. X. L. Bao, G. Ruvio, M. J. Ammann, and M. John, "A novel GPS patch antenna on a fractal hi-impedance surface substrate," *IEEE Antennas Wireless Propagat. Lett.*, **vol. 5**, 323–6, 2006.
37. P. Salonen and Y. Rahmat-Samii, "Textile antennas: effects of antenna bending on input matching and impedance bandwidth," *IEEE Aerospace Electronic Systems Magazine*, **vol. 22, no. 3**, 10–14, 2007.
38. P. Salonen, M. Keskilammi, and L. Sydanheimo, "A low-cost 2.45 GHz photonic band-gap patch antenna for wearable systems," *Proc. 11th Int. Conf. Antennas and Propagation ICAP*, pp. 719–24, April 17–20, 2001.
39. P. Salonen, F. Yang, and Y. Rahmat-Samii, "WEBGA – wearable electromagnetic band-gap antenna," *IEEE APS Int. Symp. Dig.*, vol. 1, 451–4, Monterey, CA, June 2004.
40. P. K. Kelly, L. Diaz, M. Piket-May, and L. Rumsey, "Scan blindness mitigation using photonic bandgap structure in phased arrays," *Proc. SPIE*, vol. 3464, 239–48, July 1998.
41. L. Zhang, J. A. Castaneda, and N. G. Alexopoulos, "Scan blindness free phased array design using PBG materials," *IEEE Trans. Antennas Propagat.*, **vol. 52, no. 8**, 2000–7, 2004.
42. Y. Fu and N. Yuan, "Elimination of scan blindness in phased array of microstrip patches using electromagnetic bandgap materials," *IEEE Antennas Wireless Propagat. Lett.*, **vol. 3**, 63–5, 2004.

6 Low profile wire antennas on EBG ground plane

In the preceding chapter, the surface wave band gap of EBG structures was used to enhance the performance of microstrip patch antennas. Another important property of EBG structures is the phase response to the plane wave illumination, where the reflection phase changes from 180° to −180° as the frequency increases. In this chapter, we utilize this property to improve the radiation efficiency of wire antennas near a ground plane. Consequently, a novel type of low profile antennas referred to as wire-EBG antennas is proposed [1–2]. A series of design examples are illustrated with diverse radiation characteristics.

6.1 Dipole antenna on EBG ground plane

6.1.1 Comparison of PEC, PMC, and EBG ground planes

In wireless communications, it is desirable for antennas to have a low profile configuration. In such a design, the overall height of the antenna structure is usually less than one tenth of the operating wavelength. A fundamental challenge in *low profile wire antenna design* is the coupling effect of a nearby ground plane.

To illustrate this effect, let's examine the performance of a simple dipole antenna near three different ground planes, namely, perfect electric conductor (PEC), perfect magnetic conductor (PMC), and electromagnetic band gap (EBG) ground planes, as shown in Fig. 6.1. The dipole is horizontally positioned in order to achieve a low profile configuration [3–4]. The dipole length is 0.40 $\lambda_{12\,GHz}$ and its radius is 0.005 $\lambda_{12\,GHz}$, while $\lambda_{12\,GHz}$, the free space wavelength at 12 GHz, is used as a reference length to define the physical dimensions of the antenna, ground plane, and EBG structure. A finite ground plane with 1 $\lambda_{12\,GHz} \times$ 1 $\lambda_{12\,GHz}$ size is used in the analysis. The EBG structure has the following parameters:

$$W = 0.12\ \lambda_{12\,GHz},\ g = 0.02\ \lambda_{12\,GHz},\ h = 0.04\ \lambda_{12\,GHz},\ \varepsilon_r = 2.20,$$
$$r = 0.005\ \lambda_{12\,GHz}, \tag{6.1}$$

where W is the patch width, g is the gap width, h is the substrate thickness, r is the radius of the vias, and ε_r is the substrate permittivity. The height of the dipole over the top surface of the EBG ground plane is 0.02 $\lambda_{12\,GHz}$. Thus, the overall antenna height measured from the bottom ground plane of the EBG structure is 0.06 $\lambda_{12\,GHz}$. The dipole

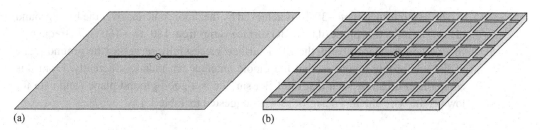

Fig. 6.1 Dipole antennas over (a) PEC or PMC ground plane and (b) EBG ground plane.

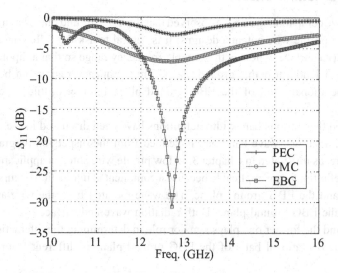

Fig. 6.2 Return loss comparison of dipole antennas over PEC, PMC, and EBG ground planes.

height over the PEC and PMC ground plane is then set to 0.06 $\lambda_{12\,\text{GHz}}$ so that all three cases have the same overall height for a fair comparison.

Figure 6.2 compares the FDTD simulated return loss of the dipole antennas over the PEC, PMC, and EBG ground plane. The input impedance is matched to a 50 Ω transmission line. When the PEC surface is used as the ground plane, the return loss of the dipole is only −3.5 dB. The reason is that the PEC surface has a 180° reflection phase and the image current has an opposite direction to that of the original dipole. The reverse image current cancels the radiation of the dipole, resulting in a very poor return loss.

When the PMC surface, which has a reflection phase of 0°, is used as the ground plane the dipole has a return loss of −7.2 dB. It is improved compared to the PEC ground plane, but it is still not desirable because a strong mutual coupling between the image current and the dipole occurs due to their close proximity. Thus, the input impedance of the dipole is changed. As a result, it cannot directly match well to a 50 Ω transmission line. Although one can use a proper impedance transformer to obtain a good return loss of the dipole, this will increase the complexity of the antenna design. Moreover, the PMC surface is an ideal surface that does not exist in nature.

The best return loss of −30 dB is achieved by the dipole antenna over the EBG ground plane. The reflection phase of the EBG surface varies from 180° to −180° with frequency. In a certain frequency band, the EBG surface successfully serves as the ground plane of the low profile dipole so that the dipole antenna can radiate efficiently. From this comparison it can be seen that the EBG surface is a good ground plane candidate for low profile wire antenna designs, as was suggested in Table 1.1.

6.1.2 Operational bandwidth selection

From the above comparison, we have observed that the EBG ground plane has a desirable feature in low profile wire antenna designs. It is natural to ask the following question: for a given EBG surface, where is the suitable frequency range so that a dipole antenna can radiate efficiently near this ground plane? This frequency region can be referred to as the operational band of the EBG ground plane for low profile wire antenna applications.

It is noticed that various bandwidth definitions have been discussed in the literature. For example, a surface wave band gap is defined from the dispersion diagram of the EBG structure, as illustrated in Chapter 3. In low profile wire antenna applications, such a band gap definition is not applicable because a complicated interaction occurs between the antenna and the EBG ground plane. The wave radiated from the wire antenna is reflected by the EBG ground plane. Both radiation wave and surface wave should be considered, and the former one plays a major role in determining the interaction effect. Therefore, the operational band of the EBG ground plane is different from previous definitions.

To identify this operational band, a straightforward approach is to characterize the performance of the wire-EBG antenna at different frequencies, as shown in Fig. 6.3a. During the FDTD simulations [5], the parameters of the EBG surface are kept the same while the dipole length is changed. Thus, the dipole antenna will resonate at different frequencies. Since the electromagnetic property of the EBG ground plane is a function of frequency, its interaction with the dipole antenna is also frequency dependent. As a result, the radiation efficiency of the antenna will vary with frequency. In the operational frequency band where the EBG ground plane provides a constructive interaction, the dipole antenna can achieve a good return loss and radiation patterns. Therefore, by observing the return loss value and radiation patterns of the dipole at different frequencies, one can find the useful operational frequency band of the EBG ground plane.

The above approach is direct, but is time consuming because multiple simulations are needed for different dipole lengths. In addition, each simulation will compute the electromagnetic fields in a relatively large FDTD volume that includes tens of EBG cells. Therefore, the total computation time is relatively long.

From the computational efficiency viewpoint, it is interesting to know if one could use the reflection phase curve of the EBG structure to identify the operational frequency band. This idea comes from the fact that the reflection phase is a good parameter to reveal the reflection property of the EBG ground plane, which significantly affects the antenna performance. Thus, the plane wave model discussed in Chapter 3 is used here

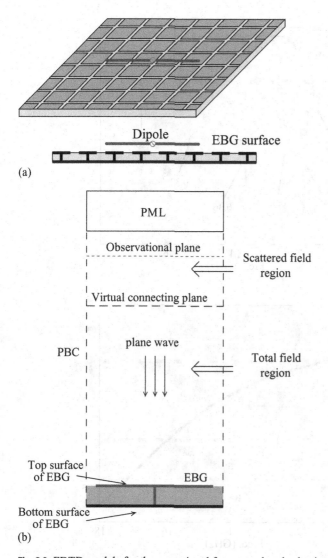

Fig. 6.3 FDTD models for the operational frequency band selection of an EBG ground plane. (a) Low profile dipole antenna over a finite EBG ground plane. (b) Plane wave normally incident upon the EBG surface.

to evaluate the reflection phase of the EBG surface, as shown in Fig. 6.3b. Since the dipole resides in the center of the ground plane, the normal incidence angle is used in the simulation as a simple approximation. With periodic boundary conditions (PBC) on four sides to model an infinite periodic structure, only a single unit of the EBG structure needs to be simulated [6–8]. Therefore, the reflection phase analysis is very efficient in computation. Finally, the frequency results of the low profile dipole model and plane wave model are compared with each other in order to establish a methodology as how to use the reflection phase curve to identify the operational frequency band.

Figure 6.4a shows the return loss results of a dipole antenna with its length varying from 0.26 $\lambda_{12\,GHz}$ to 0.60 $\lambda_{12\,GHz}$. The radius of the dipole remains 0.005 $\lambda_{12\,GHz}$. The

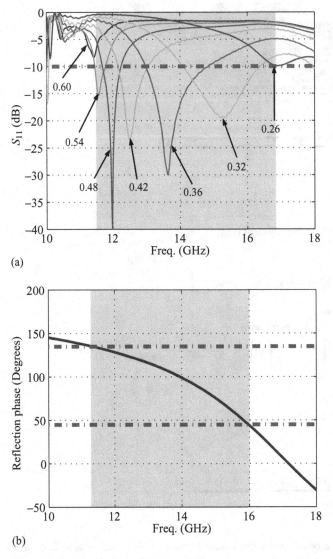

Fig. 6.4 Comparison of two FDTD model results: (a) return loss of the dipole antenna with its length varying from $0.26\, \lambda_{12\,GHz}$ to $0.60\, \lambda_{12\,GHz}$, (b) reflection phase of the EBG surface. The frequency band of the dipole model is 11.5–16.6 GHz according to $S_{11} < -10$ dB return loss criterion. The frequency band of the plane wave model is 11.3–16 GHz for $90° \pm 45°$ reflection phase region (from [4], © IEEE 2003).

height of the dipole is set to $0.02\, \lambda_{12\,GHz}$, which is only one cell above the EBG ground plane in the FDTD simulation. In order to obtain symmetrical patterns, the dipole is positioned in the center of a finite ground plane with a size of $1\, \lambda_{12\,GHz} \times 1\, \lambda_{12\,GHz}$. As the dipole length is increased, the antenna resonant frequency decreases as expected. More importantly, the value of the reflection coefficient varies with frequency. If we use

$S_{11}<-10$ dB as the criterion, we could identify a frequency band from 11.5 to 16.6 GHz (36.3%) where the dipole antenna radiates efficiently.

Figure 6.4b presents the reflection phase results of the plane wave model. If one chooses the $90°\pm45°$ reflection phases as the criterion for the EBG surface, a frequency region from 11.3 to 16 GHz can be identified, which is nearly the same frequency range obtained in the dipole model. The $90°\pm45°$ criterion is consistent with the PEC, PMC, and EBG comparison in the previous section. The PEC has a $180°$ reflection phase and the dipole antenna suffers from the reverse image current. The PMC surface has a $0°$ reflection phase and the dipole antenna does not match well due to the strong mutual coupling. When the EBG ground plane exhibits a reflection phase in the middle, a good return loss is obtained for the dipole antenna.

From this comparison, it is revealed that the reflection phase curve of an EBG surface provides a useful reference to identify its operational band for low profile wire antenna applications. The desired band is close to the frequency region where the EBG surface shows a reflection phase in the range of $90°\pm45°$. This quadrature phase is helpful for a low profile wire antenna to obtain a good return loss. One may also consider the frequency region for $-90°\pm45°$ reflection phase. It will result in a high frequency region, where the EBG cell size is relatively large in terms of the operating wavelength.

The radiation patterns of dipole antennas on the EBG surface are also computed to confirm the radiation efficiency. Figure 6.5 displays both the E- and H-plane patterns of three dipole antennas at their resonant frequencies: (1) a $0.48\ \lambda_{12\,GHz}$ dipole, which resonates at 12 GHz; (2) a $0.36\ \lambda_{12\,GHz}$ dipole, which resonates at 13.6 GHz; and (3) a $0.32\ \lambda_{12\,GHz}$ dipole, which resonates at 15.3 GHz. It is observed that all three dipoles EBG have directivities around 8 dB. The good radiation patterns also benefit from the surface wave frequency band gap of the EBG ground plane, as discussed in Chapter 5.

6.1.3 Parametric studies

To demonstrate the robustness of the quadrature phase criterion, several parametric studies are performed on the dipole antenna over the EBG ground plane. First, the relative position of the dipole with respect to the EBG ground plane is investigated. In Fig. 6.6a, there are four "+" signs labeled from A to D. Each sign indicates a center location of the dipole. The return loss results shown in Fig. 6.4a are for dipoles whose centers are located at position A, which is between EBG patches. Figure 6.6b shows the return loss results of dipoles whose centers are located at position C, which is in the center of a square patch. Although the return loss changes a bit for each dipole, the obtained operational band of the EBG ground plane is 11.5–16.5 GHz, which is very similar to the result from Fig. 6.4a. Simulations of dipoles with center positions B and D are also carried out, and the same results are observed. In addition, the ground plane size effect is also considered. A $2\ \lambda_{12\,GHz}\times 2\ \lambda_{12\,GHz}$ EBG ground plane is simulated in the dipole model and little change is noticed in the operational band. In summary, the operational band obtained from the dipole model is consistent.

The second parametric study is to verify the frequency scalability of the operational band. The analyzed EBG case has the same dielectric constant as (6.1), but the other

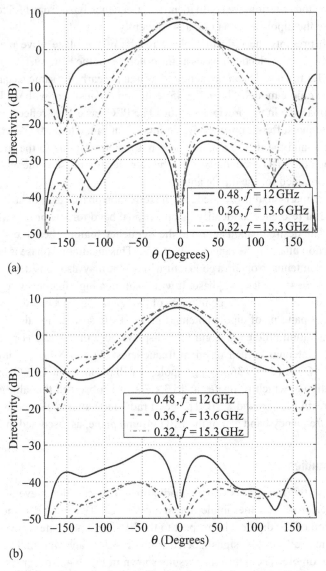

Fig. 6.5 Radiation patterns of three dipoles at their resonant frequencies. (a) E-plane patterns. (b) H-plane patterns. The 0.48 $\lambda_{12\,GHz}$ dipole resonates at 12 GHz, the 0.36 $\lambda_{12\,GHz}$ dipole resonates at 13.6 GHz, and the 0.32 $\lambda_{12\,GHz}$ dipole resonates at 15.3 GHz (from [4], © IEEE 2003).

parameters such as patch width, height, and gap width are doubled. They are listed below

$$W = 0.12\,\lambda_{6\,GHz},\ g = 0.02\,\lambda_{6\,GHz},\ h = 0.04\,\lambda_{6\,GHz},\ \varepsilon_r = 2.20,$$
$$r = 0.005\,\lambda_{6\,GHz},$$
(6.2)

where $\lambda_{6\,GHz}$ is the free space wavelength at 6 GHz. Figure 6.7 compares the reflection phase results of the EBG surface and the return loss results of dipoles with varying lengths from 0.26 $\lambda_{6\,GHz}$ to 0.60 $\lambda_{6\,GHz}$. The dipole shows a good return loss from 5.8

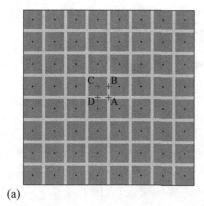

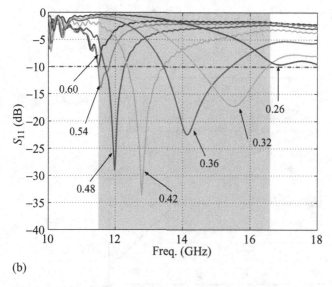

Fig. 6.6 Positional effect of the dipole on the operational band selection of the EBG ground plane: (a) various dipole positions ("+" represents the center of the dipole), (b) return loss of the dipole located at position C with its length varying from 0.26 $\lambda_{12\,GHz}$ to 0.60 $\lambda_{12\,GHz}$ (from [4], © IEEE 2003).

to 8.3 GHz, which is close to the frequency region (5.7–8.0 GHz, 33.5%) where the reflection phase of the EBG surface is in the range 90° ± 45°. The radiation patterns are calculated and the results are similar to Fig. 6.5.

The next EBG case has the same parameters as (6.1) except that the dielectric constant is increased from 2.2 to 10.2. The dipole length varies from 0.48 $\lambda_{12\,GHz}$ to 0.68 $\lambda_{12\,GHz}$, and the corresponding results are shown in Fig. 6.8. The frequency region of the reflection phase in the range 90° ± 45° is from 7.2 to 8.6 GHz (17.7%), inside which low profile dipoles exhibit good return loss results. It is worthwhile to point out that the operational bandwidth of the EBG ground plane becomes narrower because the dielectric constant is increased. These parametric studies further validate the quadrature reflection phase criterion. It can be used as a guideline for low profile wire antenna designs in the following sections.

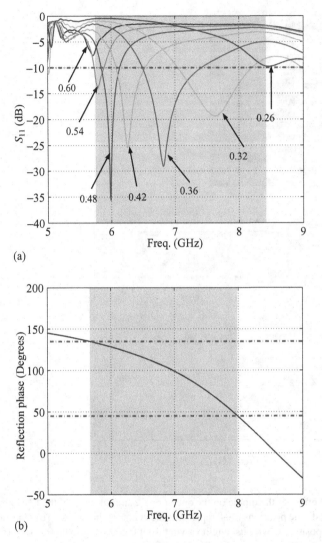

Fig. 6.7 Comparison of two FDTD model results of the EBG surface described in (6.2): (a) return loss of the dipole with its length varying from 0.26 $\lambda_{6\,GHz}$ to 0.60 $\lambda_{6\,GHz}$, (b) reflection phase of the EBG surface. The frequency band of the dipole model is 5.8–8.3 GHz according to −10 dB return loss criterion. The frequency band of the plane wave model is 5.7–8.0 GHz for 90° ± 45° reflection phase region (from [4], © IEEE 2003).

6.2 Low profile antennas: wire-EBG antenna vs. patch antenna

6.2.1 Two types of low profile antennas

Low profile antennas are preferable in modern wireless communication systems. They could be easily integrated with RF and microwave circuitry. Furthermore, they can be built conformal to installation platforms such as automobiles and vessels so that they won't affect the mechanical or aerodynamic properties of the vehicles.

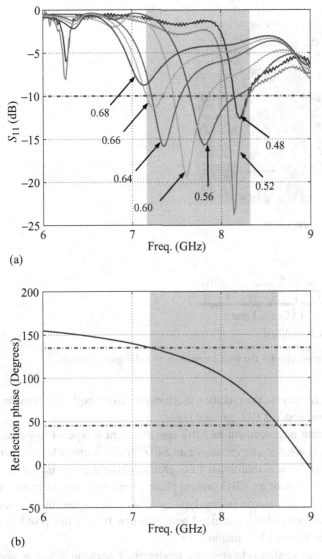

Fig. 6.8 Comparison of two FDTD model results of the EBG surface on a high permittivity substrate: (a) return loss of the dipole with its length varying from $0.48\,\lambda_{12\,\text{GHz}}$ to $0.68\,\lambda_{12\,\text{GHz}}$, (b) reflection phase of the EBG surface. The frequency band of the dipole model is 7.2–8.3 GHz according to -10 dB return loss criterion. The frequency band of the plane wave model is 7.2–8.6 GHz for $90° \pm 45°$ reflection phase region (from [4], © IEEE 2003).

Microstrip antennas are a well-known type of low profile antenna, which have gained increasing popularity in the last three decades [9–10]. Extensive research has been carried out on microstrip antennas, such as wideband approaches, miniaturization techniques, and circular polarization designs. They have been widely applied in many wireless communication systems, such as global positioning systems (GPS), wireless local area networks (WLAN), radio frequency identification (RFID) systems, etc. For a basic

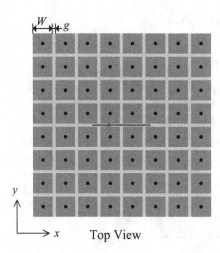

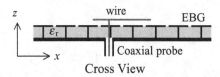

Fig. 6.9 Geometry of a probe-fed wire antenna on an EBG ground plane.

microstrip patch antenna, the radiation mechanism can be explained from an equivalent *magnetic current above a PEC ground plane*.

With the recent advancement in EBG research, a new type of low profile antenna has emerged, i.e. *electric current above an EBG ground plane*. Electric current cannot radiate efficiently near a traditional PEC ground plane because of the reverse image current. The creation of an EBG ground plane overcomes this difficulty. As a result, design of a low profile electric current type antenna becomes possible. Since the electric current is usually guided along a metal wire, this new type of low profile antenna may be referred to as a wire-EBG antenna.

The goal of this section is to answer a fundamental question in low profile antennas: what are the radiation differences between a patch antenna and a wire-EBG antenna [11]? It is believed that the answer to this question will effectively stimulate the research on the wire-EBG antenna. In addition, it can be used as a guideline to select low profile antennas in practical wireless communication systems. To address this comparative study, we'll first compare the radiation performance of these two types of low profile antennas.

6.2.2 Performance comparison between wire-EBG and patch antennas

Figure 6.9 depicts an efficient low profile wire antenna on a mushroom-like EBG ground plane. The length of the wire is 28.8 mm and it is positioned 1.6 mm above an EBG surface, which consists of 8 × 8 square patches on a thin RT/Duroid 5880 substrate.

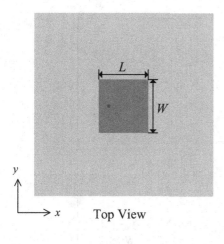

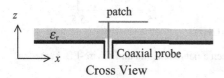

Fig. 6.10 Geometry of a probe-fed microstrip patch antenna.

The patch size is 9.6 mm, the gap width is 1.6 mm, and the patches are connected to the ground plane through center vias. The substrate has a thickness of 3.2 mm and a dielectric constant of 2.2. It is worth pointing out that the overall antenna height (4.8 mm) is much smaller than the operating wavelength (~70 mm). For the purpose of comparison, a microstrip patch antenna with the same antenna height and substrate property is designed to operate at the same frequency, as shown in Fig. 6.10. The length of the patch is 24 mm and the width of the patch is 25.6 mm. Both antennas are fed through 50Ω coaxial probes and the probe locations are adjusted for optimum return losses. In the wire-EBG antenna design, the probe is located at 8 mm from the end of the wire. In the patch antenna design, it is located 4.8 mm from the edge of the patch.

The antennas are simulated using an FDTD program [5] and the return loss results are compared in Fig. 6.11. Both antennas resonate around 4.35 GHz with good impedance match. It is noticed that the wire-EBG antenna has a broad bandwidth of 14.0% whereas the patch antenna shows a narrow bandwidth of 7.2%. This observation is intuitive from their radiation mechanisms. The patch antenna can be considered as a two-dimensional resonant cavity with a high Q factor. Thus, the antenna bandwidth is narrow. The wire antenna is also a resonant type antenna, but it is only one dimensional along the length direction. The other two dimensions are open in free space, and it is relatively difficult to store resonant energy. As a result, it has a lower Q factor and a broader operating bandwidth.

The radiation patterns of both antennas are computed and plotted in Figs. 6.12 and 6.13. The wire-EBG antenna and the patch antenna exhibit similar shape of the radiation

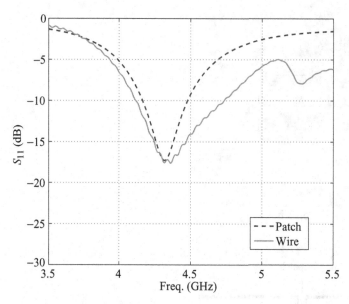

Fig. 6.11 Bandwidth comparison between a patch antenna and a wire-EBG antenna.

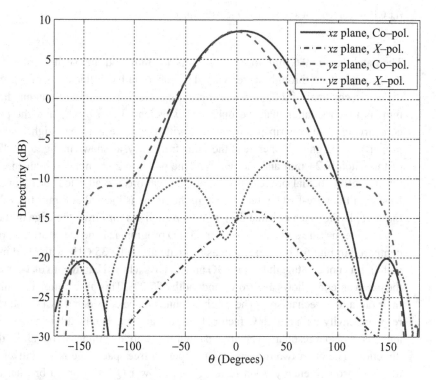

Fig. 6.12 Radiation patterns of a wire-EBG antenna at its resonant frequency of 4.35 GHz.

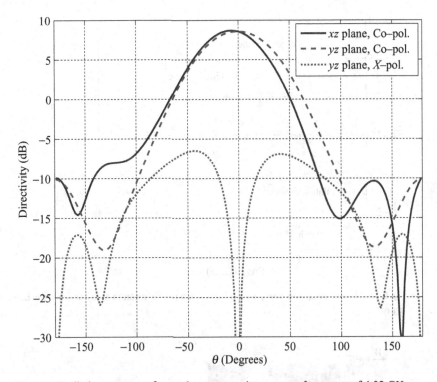

Fig. 6.13 Radiation patterns of a patch antenna at its resonant frequency of 4.32 GHz.

patterns. It is interesting to notice that the back lobes in the wire-EBG antenna are significantly lower than those of the patch antenna. The reason is that the surface wave suppression property of the EBG structure helps to reduce the edge diffraction at the boundary of the ground plane.

Furthermore, the mutual coupling in these two types of low profile antennas is investigated. Figure 6.14 shows the mutual coupling between two wire-EBG antennas as compared to the patch antennas. Both E- and H-plane couplings are calculated with the antenna distance set to 35 mm, which is half-wavelength. In general, the wire-EBG antennas have a stronger coupling than the patch antennas. At the resonant frequency of 4.35 GHz, the E-plane coupling between wire-EBG antennas is about 5 dB larger than the patch antennas while the H-plane coupling is about 3 dB larger. The consequence of this observation can be twofold. On the one hand, it may cause some problems when building an antenna array using the wire-EBG element. On the other hand, the wire-EBG antenna may be a good candidate when strong coupling is necessary, such as in a Yagi antenna design.

6.2.3 A dual band wire-EBG antenna design

As a one-dimensional geometry, one can twist the wire antenna appropriately so that the electric currents can be effectively controlled to realize diverse radiation characteristics. For example, Fig. 6.15 sketches a simple dual band wire-EBG antenna design. An

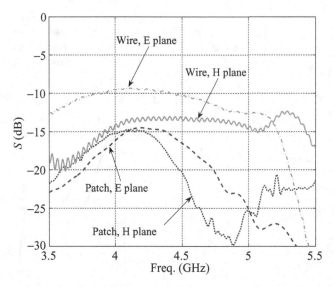

Fig. 6.14 Mutual coupling comparison between a patch antenna array and a wire-EBG antenna array.

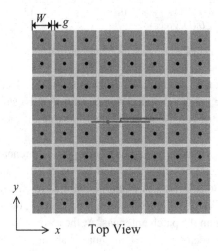

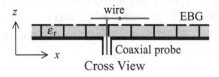

Fig. 6.15 Geometry of a dual band wire-EBG antenna.

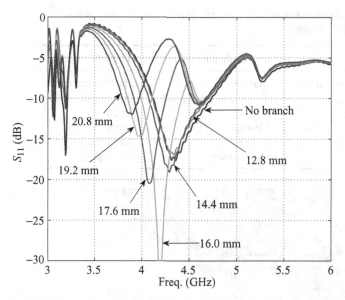

Fig. 6.16 Return losses of the dual band wire-EBG antenna with different side branch lengths.

additional wire branch is connected to the original wire at the middle point. As a result, a second resonance is introduced with the frequency determined by the length of the branch. The separation between the new branch and the original wire is set to 1.6 mm.

Figure 6.16 shows the return loss of the antenna with different lengths of the wire branch. The original wire-EBG antenna has only one resonance. As the additional wire segment is added, the overall antenna resonant behavior changes. If the length of the additional wire is similar to the original wire, still one null is observed in the return loss curves. As the length is increased, two resonances become distinguishable from each other and two nulls are observed. The higher resonant frequency remains the same, which corresponds to the original wire. The lower resonant frequency decreases with the branch length increasing. It is noticed that the frequency ratio between these two frequencies can be controlled in a rather flexible manner.

At the end of this section, it is helpful to summarize the comparison between these two types of low profile antennas. Table 6.1 lists some important differences. With the rapid development of wireless communications and the popularity in microstrip antennas, it is believed that the wire-EBG antenna with its attractive low profile configuration could become a favorable concept for the antenna engineers.

6.3 Circularly polarized curl antenna on EBG ground plane

The wire-EBG antennas discussed in previous sections are linearly polarized. In many wireless communication systems, circularly polarized electromagnetic waves are used to transmit signals, such as the GPS and satellite links. In this section, a circularly polarized curl antenna on the EBG ground plane is presented.

Table 6.1 Comparison of two types of low profile antennas: microstrip antenna versus wire-EBG antenna.

	Microstrip antenna	Wire-EBG antenna
Operation mechanism	Magnetic current	Electric current
Geometry	Two dimensional patch	One dimensional wire
Bandwidth	Narrow	Broad
Coupling	Medium	Strong
Design freedom	Flexible	Very flexible
Research	Developed	Developing

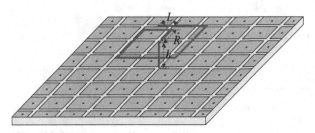

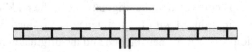

Fig. 6.17 Geometry of a square curl antenna over an EBG ground plane (from [14], © Wiley Inter Science 2001).

6.3.1 Performance of curl antennas over PEC and EBG ground planes

A curl antenna is a simple radiator to generate a circular polarization pattern [12–13]. However, it cannot function well when it is placed close to a PEC ground plane because of the reverse image current. To solve this problem, an EBG ground plane is used to replace the traditional PEC ground plane in the curl antenna designs [14].

The configuration of the curl antenna over an EBG surface is displayed in Fig. 6.17. It consists of two parts: an EBG ground plane and a square curl. The square shape is investigated in here because it is easily analyzed by the FDTD method. The parameters of the curl are the height h, inner radius R, and the extended curl length L after one round. The radius is increased by 0.1 λ at each bend.

A circularly polarized antenna is designed at GPS frequency 1.57 GHz. A finite ground plane of 1 λ × 1 λ is kept for practical applications. Here, λ is the free space wavelength at 1.57 GHz. The parameters of the EBG ground plane are scaled from (6.1). The EBG patch size is 0.12 λ × 0.12 λ and the gap between the patches is 0.02 λ. The substrate thickness is 0.04 λ with a dielectric constant of 2.20. The curl antenna is optimized to obtain good circularly polarized patterns. For the purpose of comparison, a curl antenna over a PEC ground plane is also designed. The optimized curl parameters over the PEC ground are:

$$R = 0.13\ \lambda,\ \ L = 0.11\ \lambda,\ \ h = 0.23\ \lambda. \qquad (6.3)$$

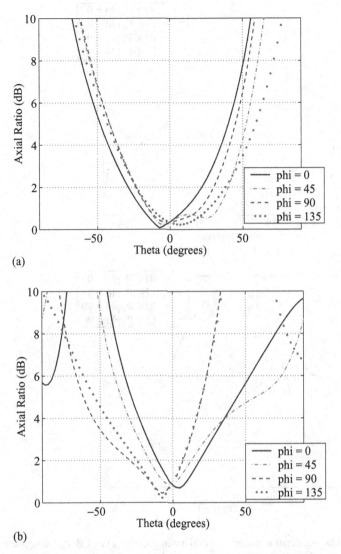

Fig. 6.18 Axial ratios of curl antennas over (a) an EBG ground plane ($h = 0.14\,\lambda$) and (b) a PEC ground plane ($h = 0.23\,\lambda$) (from [14], © Wiley Inter Science 2001).

The total length of the curl starting from the feeding probe is $1.67\,\lambda$. The optimized parameters of the curl over the EBG ground plane are:

$$R = 0.14\,\lambda, \quad L = 0.08\,\lambda; \quad h = 0.14\,\lambda. \tag{6.4}$$

In this case, the total length of the curl is $1.64\,\lambda$.

Figure 6.18 compares the simulated axial ratio of these two antennas. Both antennas have good axial ratios at the broadside. Note that the EBG case shows an attractive feature of a lower profile than the PEC case. Figure 6.19 compares the simulated circular polarization (CP) patterns of the curl antennas. Both right-hand circular polarization

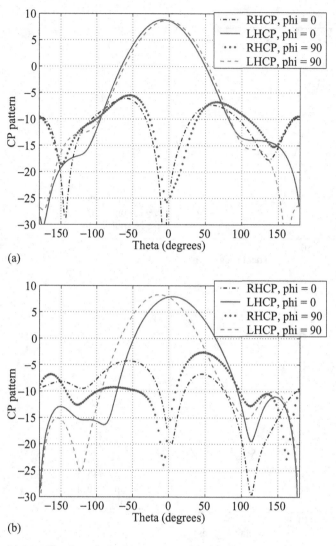

Fig. 6.19 Circular polarization patterns of curl antennas over (a) an EBG ground plane ($h = 0.14 \lambda$) and (b) a PEC ground plane ($h = 0.23 \lambda$).

(RHCP) patterns and left-hand circular polarization (LHCP) patterns are plotted, as well as the $\phi = 0°$ and $\phi = 90°$ cuts. It is observed that the EBG case has similar radiation performance at different cuts whereas the pattern changes noticeably for the PEC case.

The input impedances of two curl antennas are compared in Fig. 6.20. At the GPS frequency 1.57 GHz, the impedance of the curl over the EBG ground plane is 100-j40 Ω while the impedance of the PEC case is 200-j100 Ω. Therefore, the EBG case is easier to match with a standard 50 Ω or 75 Ω transmission line. From the comparison, it can be concluded that the EBG ground plane is also applicable in designing low profile circularly polarized wire antennas.

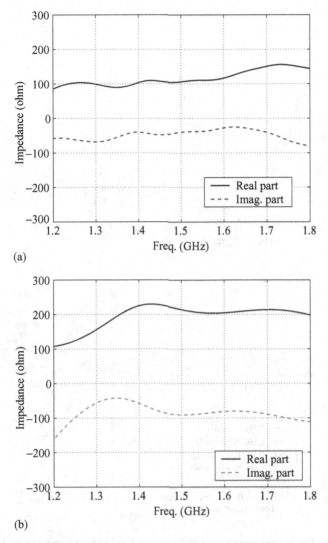

Fig. 6.20 Input impedances of curl antennas over (a) an EBG ground plane ($h = 0.14\,\lambda$) and (b) a PEC ground plane ($h = 0.23\,\lambda$).

6.3.2 Parametric studies of curl antennas over the EBG surface

To further understand the behavior of the square curl antenna over the EBG ground plane, some parametric studies, including the antenna height and ground plane size, are investigated in detail. To explore the potential of the height reduction effect of the EBG ground plane, curl antennas with decreased heights are analyzed. It is found that when the height is reduced, the axial ratio of the antenna will increase. There is a trade off between the antenna height and CP performance. For example, a lower profile curl antenna is simulated with the following parameters:

$$R = 0.14\,\lambda, \quad L = 0.13\,\lambda, \quad h = 0.06\,\lambda. \tag{6.5}$$

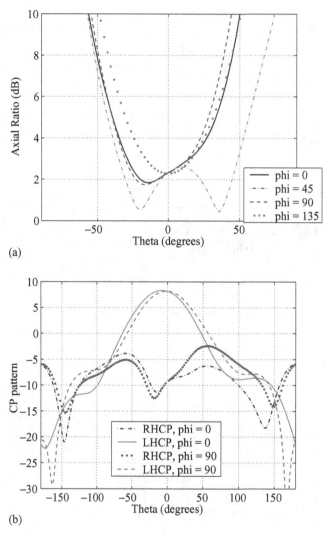

Fig. 6.21 A low profile curl antenna ($h = 0.06\ \lambda$) over an EBG ground plane: (a) axial ratio and (b) circular polarization patterns (from [14], © Wiley Inter Science 2001).

The total length of the curl is $1.61\ \lambda$. The axial ratios at different ϕ cuts are shown in Fig. 6.21a. As revealed in the figure, this antenna still generates an acceptable axial ratio less than 3dB. Figure 6.21b plots the circular polarization pattern of the curl antenna at $\phi = 0°$ and $\phi = 90°$ planes.

Ground plane size is another important factor in practical antenna applications. In the previous design, 8×8 patches on $1\ \lambda \times 1\ \lambda$ EBG ground plane are used. The size of the EBG surface is reduced to test its effect while the other parameters remain their previous values in (6.4). Two cases are analyzed:

Case 1: $0.84\ \lambda \times 0.84\ \lambda$ EBG surface, including 6×6 patches
Case 2: $0.56\ \lambda \times 0.56\ \lambda$ EBG surface, including 4×4 patches

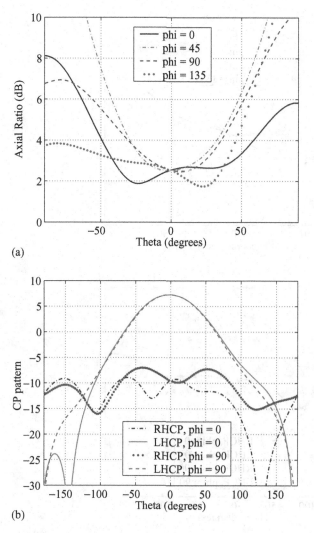

Fig. 6.22 A curl antenna over 0.84 λ × 0.84 λ EBG ground plane: (a) axial ratio and (b) circular polarization patterns.

The axial ratios of Case 1 at different ϕ cuts are shown in Fig. 6.22a. Although the ground plane size is reduced, the antenna still generates an acceptable axial ratio less than 3dB. Figure 6.22b plots the circular polarization patterns of the curl antenna at $\phi = 0°$ and $\phi = 90°$ planes. The antenna gain is about 7.5 dB and the cross-polarization at broadside is around −10 dB.

When the ground plane size is further reduced, the axial ratio becomes larger than 3 dB. Fig. 6.23a plots the axial ratio of Case 2. The CP patterns are presented in Fig. 6.23b. Since the ground plane in this case is too small, the gain is reduced to 4.9 dB while the cross-polarization is increased to −7.5 dB. Therefore, it can be summarized from this study that a minimum EBG ground plane size of 0.84 λ × 0.84 λ (including 6×6 EBG cells) is required to obtain good CP patterns.

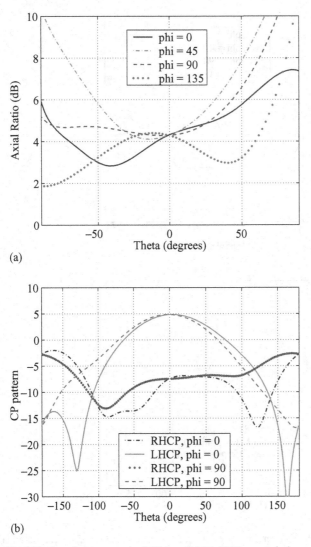

Fig. 6.23 A curl antenna over 0.56 λ × 0.56 λ EBG ground plane: (a) axial ratio and (b) circular polarization patterns.

6.3.3 Experimental demonstration

Some experiments have been performed to prove the validity of this antenna design. Since the EBG surface for 1.57 GHz is relatively large, antennas are scaled to 6 GHz for ease of fabrication and measurements.

Figure 6.24 shows a photograph of a curl antenna over an EBG ground plane. The EBG surface is built on 2 mm thick RT/duroid 5880 ($\varepsilon_r = 2.20$). The EBG patch is 6 mm × 6 mm and the gaps between patches is 1 mm wide. The patches are connected to the ground plane by vias in the center of the patches. The overall size of EBG structure is 52 mm × 52 mm. The parameters of the curl are:

$$R = 5.5 \text{ mm}, \quad L = 5 \text{ mm}, \quad h = 3 \text{ mm}. \tag{6.6}$$

Fig. 6.24 Photograph of a curl antenna over an EBG ground plane (from [14], © Wiley Inter Science 2001).

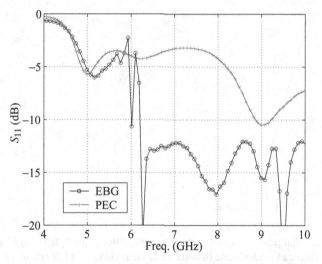

Fig. 6.25 Comparison of measured return losses of curl antennas over an EBG ground plane and a PEC ground plane (from [14], © Wiley Inter Science 2001).

Figure 6.25 compares the return loss of the curl antenna over the EBG and PEC ground planes. Since the height is only 3 mm, the curl over the PEC ground plane is not matched well. In contrast, the curl over the EBG ground plane shows an improved match in the frequency range from 6 GHz to 8.5 GHz due to the in-phase reflection feature of the EBG ground plane.

Figure 6.26a shows the measured antenna axial ratio (AR) at broadside of the antenna versus frequency. The curl antenna over the EBG surface achieves a good axial ratio (AR) of 0.9 dB at 7.18 GHz. It should be pointed out that according to Fig. 6.7b, the reflection phase of this EBG structure is around 90° at this frequency. An 8.4% CP band width (AR < 3dB) is obtained for this design. Figure 6.26b plots the measured axial ratio

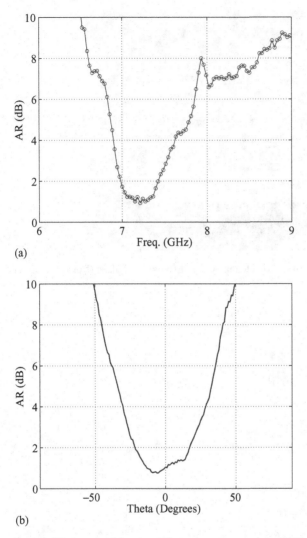

Fig. 6.26 Measured axial ratio of the curl antenna over the EBG ground plane: (a) axial ratio vs. frequency at the antenna broadside and (b) axial ratio vs. angle at 7.18 GHz (from [14], © Wiley Inter Science 2001).

versus elevation angle at frequency 7.18 GHz. It is noticed that this curl antenna has a 3 dB AR beam ranging from −25 degrees to 25 degrees.

6.4 Dipole antenna on a PDEBG ground plane for circular polarization

Besides the curl antenna on an EBG ground plane, another circularly polarized wire-EBG antenna is introduced in this section. In this design, a linearly polarized dipole antenna is located on top of a *polarization-dependent EBG (PDEBG) ground plane* [15]. The PDEBG ground plane not only improves the radiation efficiency, but also changes the linear polarization into circular polarization. The latter effect is similar to a meander-line polarizer [16].

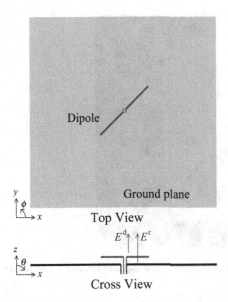

Fig. 6.27 Dipole antenna near a ground plane. The dipole is oriented along $\phi = 45°$ direction. The total radiating field consists of directly radiating field and reflected field from the ground plane (from [15], © IEEE 2005).

6.4.1 Radiation mechanism of CP dipole antenna

Figure 6.27 shows the geometry of a horizontal dipole near a ground plane. The dipole is oriented along $\phi = 45°$ direction and its height over the ground plane is very small compared to the operating wavelength (i.e. 0.02 λ). The total radiation field at the broadside direction can be approximated by the summation of the directly radiating field ($\vec{E}^d$) from the dipole and the reflected field ($\vec{E}^r$) from the ground plane as below:

$$\vec{E} = \vec{E}^d + \vec{E}^r = \frac{E_0}{2}\left(\hat{x}\cdot e^{-jkz} + \hat{y}\cdot e^{-jkz}\right)$$
$$+ \frac{E_0}{2}\left(\hat{x}\cdot e^{-jkz-2jkd+j\theta_x} + \hat{y}\cdot e^{-jkz-2jkd+j\theta_y}\right) \quad (6.7)$$

where E_0 denotes the magnitude of electric fields, k is the free-space wavenumber, d is the height of the dipole over the ground plane, $\hat{x}$ and $\hat{y}$ are unit vectors shown in Fig. 6.27, θ_x is the reflection phase of the ground plane for the x-polarized incident wave, and θ_y is for the y-polarized incident wave. When the dipole is located very close to the ground plane, $2kd$ in (6.7) is close to zero. Thus, the radiation field becomes:

$$\vec{E} = \frac{E_0}{2}\left(\hat{x}\cdot e^{-jkz} + \hat{y}\cdot e^{-jkz}\right) + \frac{E_0}{2}\left(\hat{x}\cdot e^{-jkz+j\theta_x} + \hat{y}\cdot e^{-jkz+j\theta_y}\right) \quad (6.8)$$

If the ground plane is a perfect electric conductor (PEC), $\theta_x = \theta_y = 180°$. The reflected field has the opposite sign to the incident fields. Thus, the total radiating field in (6.8) is zero. If the ground plane is a perfect magnetic conductor (PMC), $\theta_x = \theta_y = 0°$. Then, $\vec{E} = E_0 e^{-jkz}(\hat{x} + \hat{y})$. The dipole still radiates linearly polarized waves.

In order to obtain a circular polarization, different reflection phases θ_x and θ_y are needed. When a PDEBG surface with reflection phases of $\theta_x = 90°$ and $\theta_y = -90°$ is

Fig. 6.28 Photograph of a low profile dipole antenna that radiates circularly polarized (CP) patterns (from [15], © IEEE 2005).

used as the ground plane, the total field becomes:

$$\vec{E} = \frac{E_0}{2} e^{-jkz}[(\hat{x} + \hat{y}) + j(\hat{x} - \hat{y})] \tag{6.9}$$

The reflected field becomes perpendicular to the directly radiating field with a 90° phase difference. Therefore, a right-hand circularly polarized (RHCP) electromagnetic wave is obtained.

Equation (6.9) conceptually explains the radiation mechanism of the circularly polarized dipole antenna. A rigorous characterization must consider the complex interactions between the dipole and the EBG ground plane, such as the dipole height and finite size of the ground plane. Full wave analysis methods such as the finite difference time domain (FDTD) method are used in the antenna designs.

6.4.2 Experimental results

To verify this antenna concept, a PDEBG ground plane is designed using the guidelines discussed in Chapter 4. The PDEBG ground plane is fabricated on a RT/duroid 6002 high frequency laminate ($\varepsilon_r = 2.94 \pm 0.04$) with a 6.10 mm thickness. The patch dimensions and gap widths are designed to achieve a 90° reflection phase for x polarization and a −90° reflection phase for y polarization. The width of the rectangular patch is 8 mm and the length is 13 mm. The gap width between adjacent patches is 2 mm along the x direction and 1 mm along the y direction. A finite ground plane with a size of 100 mm × 100 mm is used in the experiment, which includes 9 × 6 rectangular patches.

A dipole antenna is mounted on this surface, as shown in Fig. 6.28. The 45° oriented dipole is positioned in the center of the ground plane. The length of the dipole is 34 mm, the height is 3 mm, and the wire radius is 0.34 mm. It is fed by a 50 Ω coaxial cable: one arm is soldered to the center conductor of the coax. and the other arm is soldered to the

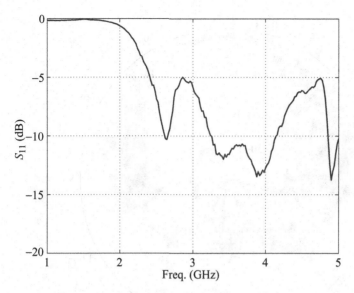

Fig. 6.29 Measured return loss of the low profile CP dipole antenna (from [15], © IEEE 2005).

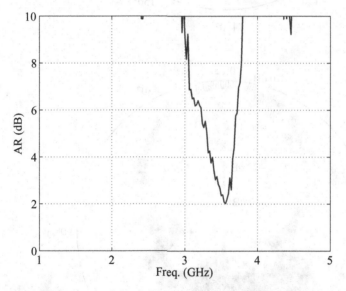

Fig. 6.30 Measured axial ratio (AR) of the antenna in the broadside direction (from [15], © IEEE 2005).

outside conductor of the coax. The outside conductor of the coax. is also soldered to the bottom conductor of the artificial ground plane.

Figure 6.29 shows the measured return loss result, which is better than −10 dB in the frequency range of 3.25–4.14 GHz. The axial ratio of the antenna at the broadside direction is measured and plotted in Fig. 6.30. An axial ratio of 2 dB is obtained at 3.56 GHz. The 3 dB axial ratio bandwidth is 200 MHz (3.45–3.65 GHz, 5.6%).

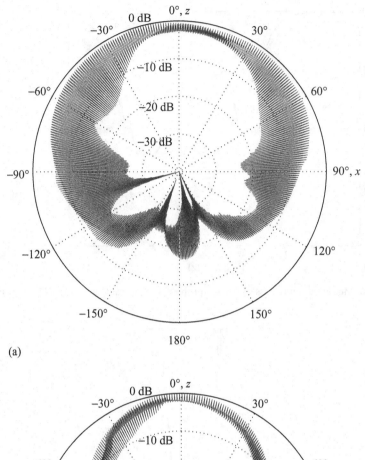

(a)

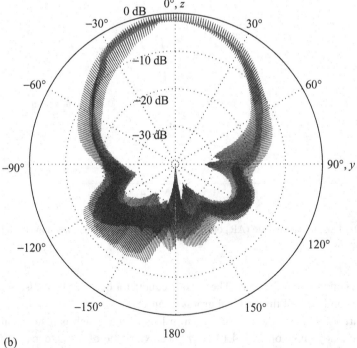

(b)

Fig. 6.31 Measured spinning linear radiation patterns of the antenna at 3.56 GHz: (a) the xz plane and (b) yz plane (from [15], © IEEE 2005).

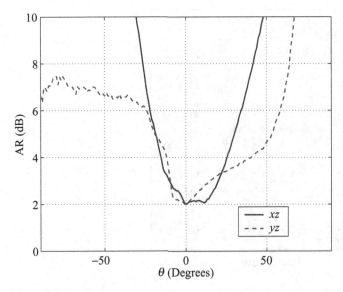

Fig. 6.32 Measured axial ratio (AR) at 3.56 GHz versus elevation angle in the xz plane and yz plane (from [15], © IEEE 2005).

The measured spinning linear patterns at 3.56 GHz are shown in Fig. 6.31. From the spinning linear pattern, the axial ratio versus elevation angle can be calculated, as plotted in Fig. 6.32. It is noticed that the 3 dB axial ratio beamwidth is around 30° in both the xz and yz plane.

It is worthwhile to point out that this antenna structure can be used to realize reconfigurable polarizations. When the orientation of the dipole is changed, the antenna can radiate electromagnetic waves with different polarizations. For example, if the dipole is oriented along the x or y direction, a linear polarization (LP) is obtained. If the dipole is oriented along $\phi = 135°$ direction, a left-hand circular polarization (LHCP) can also be achieved.

6.5 Reconfigurable bent monopole with radiation pattern diversity

Reconfigurable antennas have attracted increasing attention in modern wireless communication systems because they can provide more functionalities than classic antenna designs [17–18]. Various design specifications can be satisfied in a single antenna unit by reconfiguring its radiation characteristics such as the operating frequencies, polarizations, radiation patterns or their combinations. The reconfigurability is realized through different approaches, including mechanical tuning, semiconductor devices such as diodes or varactors, or micro-electro-mechanical system (MEMS) actuators. Compared to traditional designs, reconfigurable antennas have attractive advantages such as compact antenna volume and low co-site interference.

In this section, we introduce two wire-EBG antennas with reconfigurable radiation patterns. Radiation pattern reconfigurability is used to realize beam scanning in radar systems, and to avoid noise sources and direct signals toward intended users in wireless

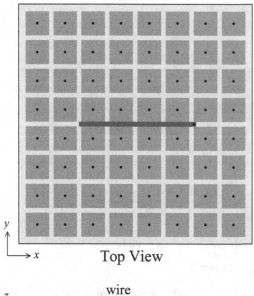

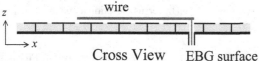

Fig. 6.33 Geometry of an *x*-oriented bent monopole on a mushroom-like electromagnetic band gap (EBG) ground plane (from [20], © John Wiley and Sons 2006).

communication networks. Although beam scanning technique has been well developed in *large* antenna systems such as phased arrays and reflector antennas, it is still a challenging area for *small* antenna elements that are popularly used in personal communication devices.

6.5.1 Bent monopole antenna on EBG ground plane

First, we study the performance of a bent monopole antenna near an EBG ground plane, as shown in Fig. 6.33. The parameters of the EBG ground plane are as follows:

$$W = 7.5 \text{ mm}, \ g = 1.5 \text{ mm}, \ h = 3 \text{ mm}, \ \varepsilon_r = 2.94, \ r = 0.375 \text{ mm} \quad (6.10)$$

These parameters guarantee that the EBG ground plane provides a suitable operation frequency around 4–5 GHz for wire antennas [19–20]. The size of the ground plane is 75 mm × 75 mm, including 8 × 8 EBG cells. The bent monopole, which is made of a 1.5 mm width strip, is located in the center of the ground plane and the feeding probe is connected to one end of the strip. To maintain the low profile advantage, the height of the bent monopole over the top ground plane is set to 1.5 mm.

When the strip length is 39 mm, the antenna resonates at 4.40 GHz with a good return loss of −15 dB and a 13.7% bandwidth, as shown in Fig. 6.34a. An interesting feature of this antenna structure is observed in its radiation patterns shown in Fig. 6.34b. In the

6.5 Reconfigurable antenna with pattern diversity

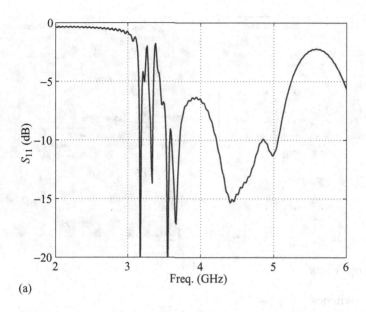

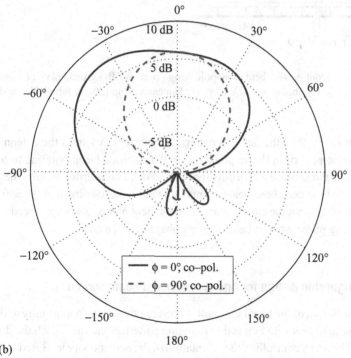

Fig. 6.34 Radiation characteristics of the bent monopole antenna on the EBG ground plane: (a) FDTD simulated return loss and (b) FDTD simulated radiation patterns (from [20], © John Wiley and Sons 2006).

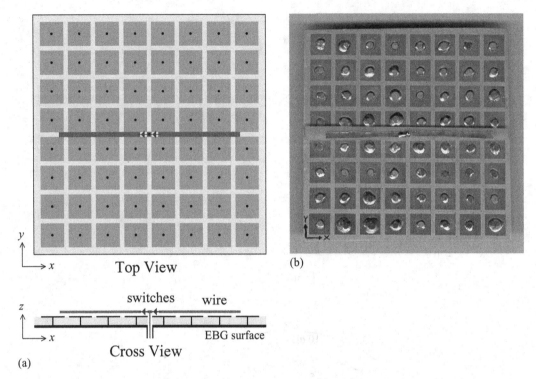

Fig. 6.35 A reconfigurable bent monopole design over an EBG ground plane: (a) antenna geometry and (b) photograph of a fabricated antenna (from [20], © John Wiley and Sons 2006).

yz plane ($\phi = 90°$) the antenna pattern is symmetric whereas the antenna exhibits an asymmetric pattern in the xz plane ($\phi = 0°$). A tilted beam pointing to $\theta = -36°$ is obtained in the xz plane with a directivity of 7.5 dB. This is different from the performance of a dipole antenna whose beam points into the broadside direction ($\theta = 0°$). The tilted beam indicates that the electric current in the bent monopole has a preceding phase near the feeding probe while it has a lagging phase near the end.

6.5.2 Reconfigurable design for one-dimensional beam switch

The tilted beam of the bent monopole structure provides an opportunity to design reconfigurable antennas with adjustable radiation patterns: one can switch the direction of an antenna beam by controlling the orientation of the bent monopole. Based on this concept, a reconfigurable wire antenna is designed, as shown in Fig. 6.35a. The same EBG ground plane is used in this structure with a feeding probe at the center. The probe is connected to two metal strips through two switches. When the left switch is ON and the right switch is OFF, the probe has an electrical connection to the left strip and the bent monopole is oriented along the $-x$ direction. When the left switch is OFF and the right switch is ON, the probe has an electrical connection to the right strip and the bent monopole is oriented along the $+x$ direction. As a result, the direction of the antenna beam can be switched and the diversity in the radiation pattern can be obtained.

To demonstrate the switchable beam concept a reconfigurable antenna prototype is built, as shown in Fig. 6.35b. A RT/duroid 6002 high frequency laminate ($\varepsilon_r = 2.94 \pm 0.04$) with 120 mil (3.048 mm) thickness is used as the substrate and the ground plane dimensions are the same as those given in (6.10). The length of the strip is tuned to 30 mm to obtain the similar resonant frequency as the single bent monopole antenna. The adjustment of the dipole length may come from the coupling effect from the parasitic strip. The feeding probe is alternatively soldered to the left and right metal strips to represent the operational mode of the $-x$ and $+x$ oriented bent monopoles. The validity of this ON/OFF representation of the switch status has been verified in [21].

The measured return loss result of the antenna is shown in Fig. 6.36a. The antenna resonates at 4.40 GHz with a good return loss of -20 dB. The bandwidth of the antenna ($S_{11} < -10$ dB) is 8.2%, similar to the FDTD simulation result. It is worthwhile to point out that the return loss of the antenna remains the same regardless of the $+x$ or $-x$ orientation of the bent monopole. A secondary resonance near 5 GHz is also noticed, which is lower than the FDTD simulation result due to fabrication errors such as the bumpy EBG surface. Our focus here is on the radiation pattern diversity at the first resonant frequency.

The radiation patterns of the antenna are measured at the resonant frequency 4.40 GHz, and Fig. 6.36b shows the measured data normalized to the maximum antenna gain. A switchable antenna beam is observed in the xz plane (E plane). When the bent monopole was $-x$ oriented, the antenna beam pointed to $\theta = 26°$ with a gain of 6.5 dB. When the bent monopole was $+x$ oriented, the antenna beam was switched to $\theta = -26°$ with the same gain. It is also noticed that beam direction in this reconfigurable wire design is opposite to that of a single bent monopole, which is contributed to by the parasitic strip that acts as a director to change the beam direction. The cross polarization level is -20 dB below the co-polarization in the xz plane and -10 dB below in the yz plane. The measured results validate the concept of this radiation pattern reconfigurable antenna.

In a practical reconfigurable antenna, a biasing circuit that controls the status of switches is an important design issue. The switches can be either pin diodes or MEMS actuators. Here, a biasing technique for pin diodes is proposed, which is also a useful reference for MEMS switches. To maintain a zero DC voltage in the strips, shorting pins are used to connect the left and right metallic strips to the ground plane. Different positions of shorting pins are calculated and the optimum location is in the center of the strips. The simulation results show that the shorting pins in this location have minimum effect on the radiation performance of the antenna. Thus, the return loss and radiation patterns are almost the same as the antenna without pins. A DC biasing voltage is supplied from the probe to control the switches. When a positive voltage is supplied, the left switch is ON and the right switch is OFF. The bent monopole orients towards the $-x$ direction. When a negative voltage is supplied, the left switch is OFF and the right switch is ON. The bent monopole orientation is switched to the $+x$ direction. The advantage of this biasing technique is that no additional DC circuits are needed and the radiating performance of the antenna remains the same.

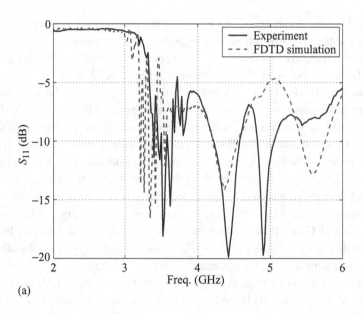

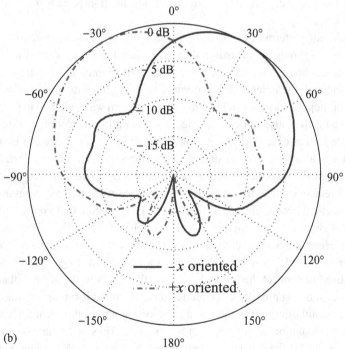

Fig. 6.36 Measured results of the reconfigurable bent monopole antenna: (a) return loss results, (b) normalized E plane radiation patterns at 4.40 GHz. Switchable beams between ±26° are observed (from [20], © John Wiley and Sons 2006).

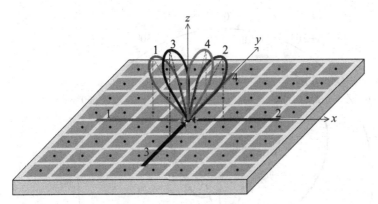

Fig. 6.37 A reconfigurable antenna scheme with two-dimensional radiation pattern diversity (from [20], © John Wiley and Sons 2006).

6.5.3 Reconfigurable design for two-dimensional beam switch

The reconfigurable concept is also extended to realize two-dimensional (2-D) beam switching, as shown in Fig. 6.37. Four strips are connected to the center probe through four switches. By controlling the switches, the bent monopole can orient along the $-x, +x, -y, +y$ directions, respectively, resulting in four different antenna beams. Therefore, the antenna beam can be switched not only in the xz plane, but also in the yz plane.

Figure 6.38 presents the simulated radiation patterns of the reconfigurable antenna in two operational modes: (1) switch 1 ON and all others OFF, (2) switch 4 ON and all others OFF. When only the switch 1 is ON, the antenna orients along the $-x$ direction and the antenna beam points to $\theta = -35°$ in the xz plane. The y-oriented strips 3 and 4 act as reflectors and isolate the coupling between strips 1 and 2. Thus the antenna beam returns back to the same direction as the single bent monopole. The isolation effect has been further tested by removing strip 2, and the antenna performance remains similar. The beam direction can be denoted by an angle pair such as $(-35°, 0°)$, where the first number indicates the elevation angle and the second refers to the azimuth angle. When switch 4 is ON and all other switches are OFF, the antenna beam points to $(35°, 90°)$. Thus, by controlling the status of these four switches, two-dimensional beam switching is realized.

6.6 Printed dipole antenna with a semi-EBG ground plane

Until now, all wire-EBG designs focus on geometries where wire antennas are positioned above a ground plane as shown in Fig. 6.39a. In some wireless communication systems [22], it is necessary that a dipole antenna needs to work efficiently near the edge of a ground plane, as shown in Fig. 6.39b. A representative example of this geometry is internal antenna designs for the wireless local area network (WLAN) service of a notebook computer, where the antenna is located in the narrow rim near the metal back LCD screen [23].

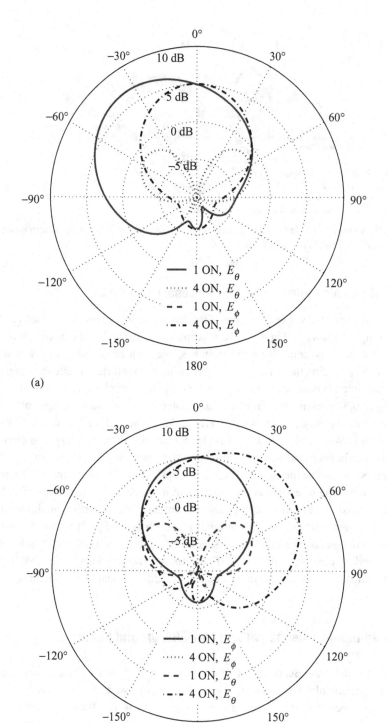

Fig. 6.38 FDTD simulated radiation patterns of the reconfigurable antenna: (a) xz plane and (b) yz plane. Two-dimensional beam switching is realized by controlling the switch status.

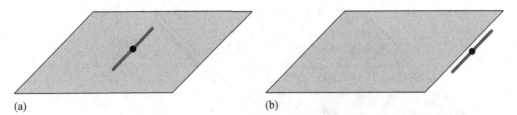

(a)　　　　　　　　　　　　　(b)

Fig. 6.39 Geometries of (a) a dipole antenna above a ground plane and (b) a dipole antenna near the edge of a ground plane (from [24], © The Electromagnetic Academy 2006).

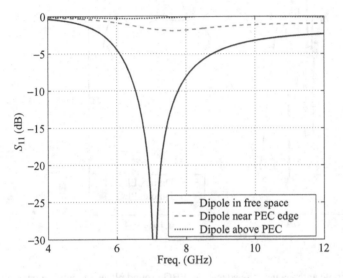

Fig. 6.40 Comparison of dipole performance in different environments. When a dipole is placed close to the edge of a conventional PEC ground plane, it cannot radiate efficiently (from [24], © The Electromagnetic Academy 2006).

6.6.1 Dipole antenna near the edge of a PEC ground plane

A half-wavelength dipole antenna radiates efficiently in free space. When it is placed around a PEC ground plane, surface currents are induced on the ground plane. When the dipole is located very close to the PEC ground plane, either on top or near the edge, the radiation of the induced currents will cancel the radiation from the dipole, resulting in low radiation efficiency.

To illustrate this effect, the FDTD program is used to simulate the performance of dipole antennas in three configurations: dipole in free space, dipole above a PEC ground plane, and dipole near the edge of a PEC ground plane. A strip dipole is used in the simulation with a length of 0.50 $\lambda_{8\,GHz}$, and width of 0.02 $\lambda_{8\,GHz}$, where $\lambda_{8\,GHz}$ is the free space wavelength at 8 GHz. A finite PEC ground plane with a size of 1 $\lambda_{8\,GHz} \times 1\ \lambda_{8\,GHz}$ is used in the simulation. The distance between the dipole and the PEC ground plane is 0.02 $\lambda_{8\,GHz}$. A gap source [5] is used to feed the dipole in the simulations.

It is observed in Fig. 6.40 that the dipole achieves a good return loss in free space. When the dipole is placed above a PEC ground plane, the return loss is close to zero, which means that very little energy is radiated in this situation. The reason is that the

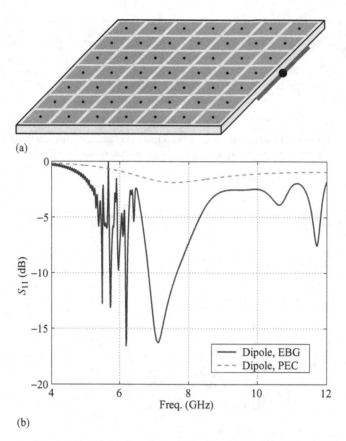

Fig. 6.41 A dipole antenna near the edge of an EBG ground plane: (a) antenna geometry and (b) FDTD simulated return loss results (from [24], © The Electromagnetic Academy 2006).

radiation from the reverse image current cancels the radiation from the original dipole. When the dipole is put near the edge of the PEC ground plane, the return loss is better than the previous case because there is not a complete reverse image current. The radiation from induced currents on the ground plane, however, still cancels most of the radiation from the dipole because the dipole is very close to the ground plane. Thus, the return loss is poor as well, which is only −2 dB.

6.6.2 Enhanced performance of dipole antenna near the edge of an EBG ground plane

The performance of a dipole antenna above a ground plane has been improved using an EBG surface in Section 6.1. It is the goal of this section to investigate the performance improvement of a dipole near the edge of a ground plane, as shown in Fig. 6.39b [24].

For this purpose, the mushroom-like EBG ground plane is again used to replace the conventional PEC ground plane, as shown in Fig. 6.41a. Periodic square patches are mounted on a grounded dielectric slab with 0.04 $\lambda_{8\ GHz}$ thickness and a dielectric constant of 2.94. The width of periodic patches is 0.12 $\lambda_{8\ GHz}$ and the gap width between adjacent patches is 0.02 $\lambda_{8\ GHz}$. Center vias with 0.005 $\lambda_{8\ GHz}$ radius are used to connect the patches to the bottom PEC conductor. Basically, these dimensions are scaled from (6.1) except that the dielectric constant is increased a little due to available substrate materials. The dipole dimensions remain as the previous dimensions.

Figure 6.41b compares the return losses of dipole antennas near the edge of an EBG ground plane and near the edge of a PEC ground plane. It is observed that the return loss of the dipole has been significantly improved from -2 dB to -16 dB. This result demonstrates the ability of the EBG ground plane to enhance the radiation efficiency of a nearby dipole.

Some parametric studies have been performed to further understand the antenna performance. Figure 6.42 depicts the effect of dipole height selections. When the dipole is located on the same plane as the top periodic patches of the EBG structure, the height is 0.04 $\lambda_{8\,GHz}$. The height is zero when the dipole is located on the same plane as the bottom conductor of the EBG structure. A negative height means that the dipole is below the bottom conductor of the EBG structure.

When the dipole height is 0, the return loss is around -4 dB. Both the EBG structure and PEC surface affect the dipole radiation and limited improvement is obtained. When the dipole height is increased, the EBG structure plays a dominant role in affecting the dipole radiation. Thus the return loss is improved significantly. A -27 dB return loss is obtained when the height is 0.06 $\lambda_{8\,GHz}$.

Several dipole positions below the bottom conductor are also simulated and the data are plotted in Fig. 6.42b. In this situation, the PEC plays a dominant role in determining the antenna efficiency. Therefore, the return loss is not as remarkably improved as in Fig. 6.42a. When the height decreases, only a slight enhancement is noticed because of the increasing distance between the dipole and the PEC ground plane.

Figure 6.43 reveals the dipole length effect on the antenna performance. In this study, the dipole is located on the same plane as the top periodic patches of EBG structures (height = 0.04 $\lambda_{8\,GHz}$). When the dipole length is increased, the resonant frequency decreases. Meanwhile, the return loss value and the antenna bandwidth change as well. It is noticed that when the resonant frequency of the antenna falls in the range of 6.5–8.3GHz, the dipole can obtain a good return loss better than -10 dB. Similar to Section 6.1, this frequency range can be defined as the operating band of the EBG structure to work as the ground plane for a nearby planar dipole antenna.

6.6.3 Printed dipole antenna with a semi-EBG ground plane

In the previous numerical study, an ideal gap source is used to excite the dipole antenna. In practical applications, the feed structure needs to be carefully selected. A microstrip fed printed dipole design is presented here, which is easy to fabricate and test experimentally.

Figure 6.44 shows photographs of a printed antenna prototype, including the front view, back view, and the final product with a 50Ω SMA connector. The antenna is built on a 60 mil (1.524mm) thick RT/Duroid 6002 substrate ($\varepsilon_r = 2.94 \pm 0.04$). The length of each dipole arm is 6.75mm and the width is 0.75 mm. The gap between the two arms is

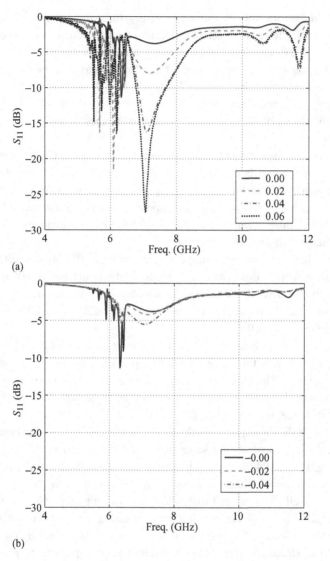

Fig. 6.42 Parametric study of the dipole height effect (Unit: $\lambda_{8\,\text{GHz}}$). When the dipole is located on the same plane as the bottom conductor of the EBG surface, the height is zero. The height is 0.04 when the dipole is located on the same plane as the top periodic patches of the EBG surface (from [24], © The Electromagnetic Academy 2006).

0.75 mm. A 50 Ω microstrip line is used to feed the dipole in order to obtain a conformal design. One arm of the dipole is directly connected to the microstrip feed line while the other arm is connected to the bottom PEC conductor through a via. A semi-EBG ground plane is used to improve the dipole performance and its dimensions are the same as described before. It is worthwhile to point out that the distance between the printed dipole and the semi-EBG ground plane is only 0.75 mm, which satisfies the compact size requirement in many wireless communication systems.

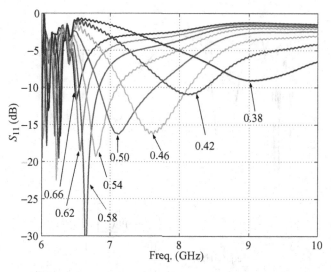

Fig. 6.43 Parametric study of the dipole antenna length effect (Unit: $\lambda_{8\,\text{GHz}}$). The return losses are calculated using the FDTD method (from [24], © The Electromagnetic Academy 2006).

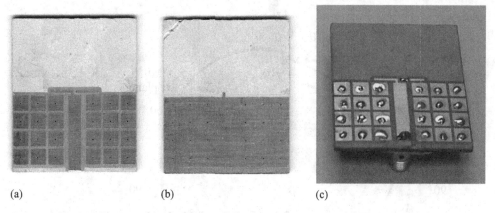

(a)   (b)   (c)

Fig. 6.44 Photographs of a fabricated dipole antenna near a semi-EBG ground plane. (a) Front view, (b) back view, and (c) final antenna prototype with an SMA connector (from [24], © The Electromagnetic Academy 2006).

Figure 6.45 presents the measured return loss results of the printed dipole antenna with the semi-EBG ground plane. A dipole antenna with the PEC ground plane is also fabricated and measured as a reference. Because of the close proximity of the dipole and the PEC ground plane, the dipole in the PEC case cannot match well to -10 dB. When the semi-EBG ground plane is used, the return loss improves significantly. The antenna resonates at 7.84 GHz with a good return loss of -20 dB. A 9.4% impedance bandwidth ($S_{11} < -10$ dB) is achieved in this design.

The antenna patterns are measured at 8 GHz and the *xz*, *yz*, and *xy* plane patterns are sketched in Fig. 6.46. It is observed that complex interaction occurs between the dipole and the EBG ground plane. The total radiation pattern is contributed to by the direct

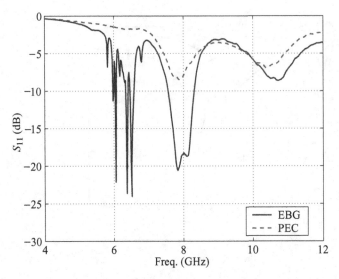

Fig. 6.45 Measured return losses of the printed dipole antennas. When the semi-EBG ground plane is used, the dipole antenna resonates at 7.84 GHz with a 9.4% bandwidth (from [24], © The Electromagnetic Academy 2006).

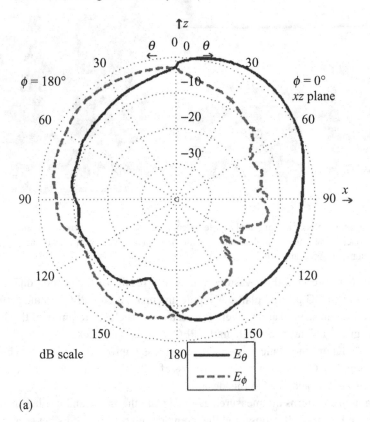

(a)

Fig. 6.46 Measured antenna patterns: (a) xz plane pattern, (b) yz plane pattern, and (c) xy plane pattern (from [24], © The Electromagnetic Academy 2006).

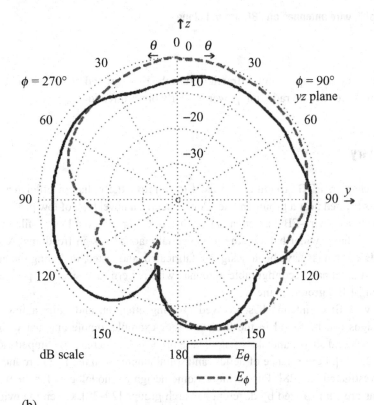

(b)

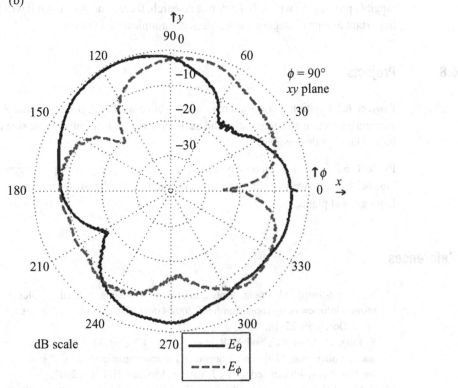

(c)

Fig. 6.46 (cont.)

radiation from the dipole and diffraction from the EBG ground plane. It can be inferred from Fig. 6.46 that the radiation power is nearly omni-directional.

6.7 Summary

In this chapter, EBG structures are used to improve the radiation efficiency of wire antennas located near a ground plane. Consequently, a novel type of low profile antenna referred to as a wire-EBG antenna is introduced. Compared to a low profile microstrip antenna, it has a wider bandwidth and potentially added design freedoms. A series of wire-EBG antennas have been designed, fabricated, and tested, including the circularly polarized antennas, reconfigurable antennas with pattern diversity, and printed dipole with semi-EBG ground plane.

The wire-EBG antenna has received growing attention and quite a few interesting designs can be found in the literature. For example, dipole antenna with bowtie shape, diamond shape, and sleeve dipole are proposed to increase the impedance bandwidth [25–26]. Performance of helical and spiral antennas on the EBG ground plane is also investigated [27–28]. Furthermore, some design methodologies for the wire-EBG antennas are summarized by different research groups [29–30]. Given the evident and tangible progress in wire-EBG antenna research, there is no doubt that it will become an important antenna category for wireless communication systems.

6.8 Projects

Project 6.1 Position a Yagi antenna array above an EBG ground plane. Analyze the antenna performance and identify the beam direction. The ground plane size needs to be investigated in this project.

Project 6.2 Design a circularly polarized antenna array using the curl design in Fig. 6.24. Compare the gain and mutual coupling of the array with and without the EBG ground plane.

References

1. Y. Rahmat-Samii and F. Yang, "Antenna developments using artificial complex ground planes: insights into new design opportunities," *2004 Asia-Pacific Microwave Conference*, New Delhi, India, December 15–18, 2004.
2. F. Yang, H. Mosallaei, and Y. Rahmat-Samii, "Chapter 34: Low Profile Antenna Performance Enhancement Utilizing Engineered Electromagnetic Materials," in *Antenna Engineering Handbook*, 4th edn., edited by J. Volakis, McGraw-Hill Inc., 2007.
3. Z. Li and Y. Rahmat-Samii, "PBG, PMC and PEC surface for antenna applications: A comparative study," *2000 IEEE APS Dig.*, pp. 674–7, July 2000.

4. F. Yang and Y. Rahmat-Samii, "Reflection phase characterizations of the EBG ground plane for low profile wire antenna applications," *IEEE Trans. Antennas Propagat.*, **vol. 51, no. 10**, 2691–703, 2003.
5. M. A. Jensen, *Time-Domain Finite-Difference Methods in Electromagnetics: Application to Personal Communication*, Ph.D. dissertation at University of California, Los Angeles, 1994.
6. H. Mosallaei and Y. Rahmat-Samii, "Periodic bandgap and effective dielectric materials in electromagnetics: characterization and applications in nanocavities and waveguides," *IEEE Trans. Antennas Propagat.*, **vol. 51, no. 3**, 549–63, 2003.
7. A. Aminian, F. Yang, and Y. Rahmat-Samii, "Bandwidth determination for soft and hard ground planes by spectral FDTD: a unified approach in visible and surface wave regions," *IEEE Trans. Antennas Propagat.*, **vol. 53, no. 1**, 18–28, 2005.
8. F. Yang, J. Chen, Q. Rui, and A. Elsherbeni, "A simple and efficient FDTD/PBC algorithm for periodic structure analysis," *Radio Sci.*, **vol. 42, no. 4**, RS4004, 2007.
9. J. J. Bahl and P. Bhartia, *Microstrip Antennas*, Artech House, 1980.
10. P. Bhartia, Inder Bahl, R. Garg, and A. Ittipiboon, *Microstrip Antenna Design Handbook*, Artech House, 2000.
11. F. Yang and Y. Rahmat-Samii, "Wire antenna on an EBG ground plane vs. patch antenna: a comparative study on low profile antennas," *URSI Electromagnetic Theory Symposium*, Ottawa, Canada, July 26–28, 2007.
12. H. Nakano, S. Okuzawa, K. Ohishi, H. Mimaki, and J. Yamauchi, "A curl antenna," *IEEE Trans. Antennas Propagat.*, **vol. 41**, 1570–5, 1993.
13. J. S. Colburn and Y. Rahmat-Samii, "Quadrifilar-curl antenna for the 'big-LEO' mobile satellite service system," *1996 IEEE APS Dig.*, pp. 1088–91, July 1996.
14. F. Yang and Y. Rahmat-Samii, "A low profile circularly polarized curl antenna over electromagnetic band-gap (EBG) surface," *Microwave Optical Tech. Lett.*, **vol. 31, no. 3**, 165–8, 2001.
15. F. Yang and Y. Rahmat-Samii, "A low profile single dipole antenna radiating circularly polarized waves," *IEEE Trans. Antennas Propagat.*, **vol. 53, no. 9**, 3083–6, 2005.
16. B. A. Munk, *Finite Antenna Arrays and FSS*, John Wiley & Sons, 2003.
17. Y. Qian and T. Itoh, "Progress in active integrated antennas and their applications," *IEEE Trans. Microwave Theory Tech.*, **vol. 46, no. 11**, 1891–900, 1998.
18. F. Yang and Y. Rahmat-Samii, "Patch antennas with switchable slots (PASS) in wireless communications: concepts, designs, and applications," *IEEE Antennas Propagat. Mag.*, **vol. 47, no. 2**, 13–29, 2005.
19. F. Yang and Y. Rahmat-Samii, "Bent monopole antennas on EBG ground plane with reconfigurable radiation patterns," *2004 IEEE APS Int. Symp. Dig.*, vol. 2, pp. 1819–22, Monterey, CA, June 20–26, 2004.
20. Y. Rahmat-Samii and F. Yang, "Chapter 12: Development of complex artificial ground planes in antenna engineering," in *Metamaterials: Physics and Engineering Explorations*, edited by N. Engheta and R. Ziolkowski, John Wiley & Sons Inc., 2006.
21. F. Yang and Y. Rahmat-Samii, "Patch antenna with switchable slot (PASS): dual frequency operation," *Microwave Optical Tech. Lett.*, **vol. 31, no. 3**, 165–8, 2001.
22. C. M. Allen, A. Z. Elsherbeni, C. E. Smith, C-W P. Huang, and K. F. Lee, "Tapered meander slot antenna for dual band personal wireless communication systems," *Microwave Optical Tech. Lett.*, **vol. 36, no. 5**, 381–5, 2003.
23. "Display located antenna specifications," *DELL Specification No. X1579*, June 2003.

24. F. Yang, V. Demir, D. Elsherbeni, A. Elsherbeni, and A. Eldek, "Enhancement of printed dipole antennas characteristics using semi-EBG ground plane," *J. Electromagnetic Waves and Applications*, **vol. 20, no. 8**, 993–1006, 2006.
25. M. G. Bray and D. H. Werner, "A broadband open-sleeve dipole antenna mounted above a tunable EBG AMC ground plane," *IEEE APS Dis.*, vol. 2, pp. 1147–50, 20–25 June 2004.
26. L. Akhoondzadeh-Asl, D. J. Kern, P. S. Hall, and D. H. Werner, "Wideband dipole on electromagnetic bandgap ground plane," *IEEE Trans. Antennas Propagat.*, **vol. 55, no. 9**, 2426–34, 2007.
27. J. M. Bell and M. F. Iskander, "A low-profile Archimedean spiral antenna using an EBG ground plane," *Antennas Wireless Propag. Lett.*, **vol. 3, no. 1**, 223–6, 2004.
28. H. Nakano, K. Hitosugi, N. Tatsuzawa, D. Togashi, H. Mimaki, and J. Yamauchi, "Effects on the radiation characteristics of using a corrugated reflector with a helical antenna and an electromagnetic band-gap reflector with a spiral antenna," *IEEE Trans. Antennas Propag.*, **vol. 53, no. 1, part 1**, 191–9, 2005.
29. S. Clavijo, R. E. Diaz, and W. E. McKinzie, "Design methodology for Sievenpiper high-impedance surfaces: an artificial magnetic conductor for positive gain electrically small antennas," *IEEE Trans. Antennas Propagat.*, **vol. 51, no. 10**, 2678–90, 2003.
30. M. F. Abedin and M. Ali, "Effects of EBG reflection phase profiles on the input impedance and bandwidth of ultrathin directional dipoles," *IEEE Trans. Antennas Propag.*, **vol. 53, no. 11**, 3664–72, 2005.

7 Surface wave antennas

The concept of surface wave antennas (SWA) was initiated in the 1950s [1–2] and numerous theoretical and experimental investigations have been reported in the literature [3–10]. To support the propagation of surface waves, a commonly used structure in SWA designs is a corrugated metal surface. However, the corrugated structure is thick, heavy, and costly, which may limit the applications of surface wave antennas in wireless communication systems.

In this chapter, novel surface wave antennas are presented. Compared to traditional SWA designs, surface waves are now guided along a *thin* grounded slab loaded with periodic patches, resulting in a low profile conformal geometry. In contrast to the previous wire-EBG antennas or patch antennas that radiate to the broadside direction, the proposed SWA achieve a monopole-like radiation pattern with a null in the broadside direction. The low profile SWA is more attractive than a traditional monopole antenna that is a quarter-wavelength high.

7.1 A grounded slab loaded with periodic patches

7.1.1 Comparison of two artificial ground planes

We start with analyzing a complex artificial ground plane, which will be subsequently used in surface wave antenna designs. Figure 7.1 shows two artificial surfaces: a mushroom-like EBG surface and a grounded dielectric slab loaded with periodic patches. In the latter structure vertical vias are removed, which results in different surface wave properties in the two ground planes.

To compare the electromagnetic properties of these two structures, the finite difference time domain (FDTD) method is used to simulate their performance [11–12]. Two electromagnetic properties of these structures are of special interest: the surface wave and the plane wave properties. The dimensions of the analyzed surfaces are:

$$W = 0.10\,\lambda, \quad g = 0.02\,\lambda, \quad h = 0.04\,\lambda, \quad \varepsilon_r = 2.94 \tag{7.1}$$

where W is the width of the square patch, g is the gap width, h is the substrate thickness and ε_r is the dielectric constant of the substrate. The vias' radius in the EBG structure is 0.005 λ, where $\lambda = 75$ mm, the free space wavelength at 4 GHz, is used as a reference length to define the physical dimensions of artificial surfaces and

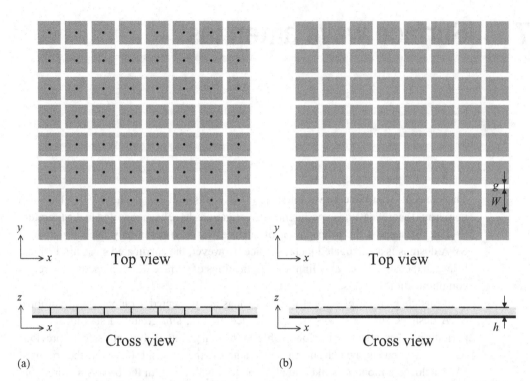

Fig. 7.1 Geometry of two artificial ground planes: (a) a mushroom-like electromagnetic band gap (EBG) structure and (b) a grounded dielectric slab loaded with periodic patches.

antennas studied in this chapter. The selection of these parameters follows the guidelines in Chapter 4.

Figure 7.2 shows the dispersion diagrams of the two artificial ground planes. Each point in the dispersion diagram represents a certain surface wave mode, where the horizontal coordinate denotes the wavenumber and the vertical coordinate represents the mode frequency. For the mushroom-like EBG structure, a frequency band gap is observed between 3.5 GHz and 5.9 GHz, which means that the surface waves inside this frequency range will be suppressed. In contrast, since the vertical vias are removed in the patch-loaded grounded slab, the band gap disappears and the first surface wave mode (TM^z dominant) can exist in the above frequency range.

The reflection phases of these two ground planes are compared in Fig. 7.3. Here, the incident plane wave is set to normally illuminate the ground planes. It is noticed that the reflection phases change continuously from 180° to −180° as frequency increases. A 90° reflection phase is achieved around 4.6 GHz and a 0° reflection phase is realized around 5.8 GHz. It is observed that removing the center vias has little effect on the reflection phase feature and both surfaces have very similar reflection phase characteristics for the normal incidence.

In summary, the patch-loaded grounded slab has a similar reflection phase as the mushroom-like EBG surface for normally incident plane waves, but it does not have a frequency band gap for surface waves.

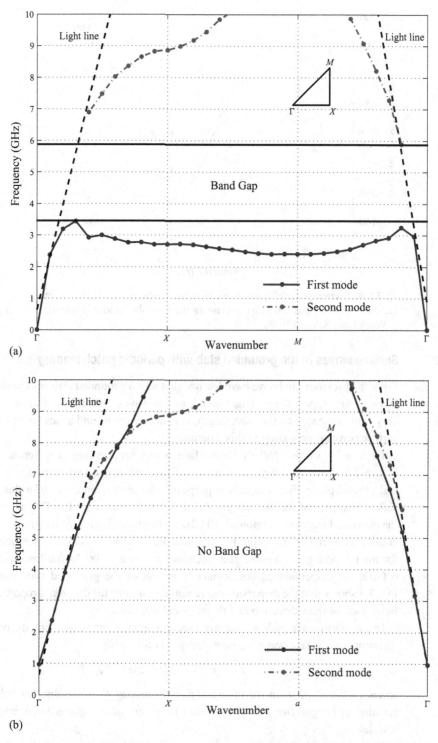

Fig. 7.2 Dispersion diagrams of two artificial ground planes: (a) a mushroom-like EBG structure and (b) a patch-loaded grounded slab. Along the horizontal axis, π: $k_x = 0$, $k_y = 0$; X: $k_x = \pi/a$, $k_y = 0$; M: $k_x = \pi/a$, $k_y = \pi/a$; $a = W + g$ (from [15], © Wiley Inter-Science 2005).

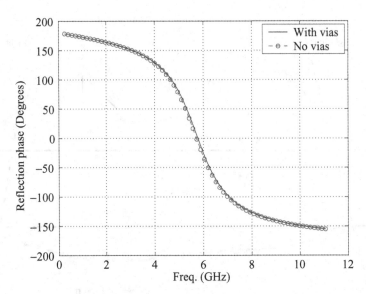

Fig. 7.3 Reflection phase characterizations of the mushroom-like EBG structure (with vias) and the patch-loaded grounded slab (no vias) for the normally incident plane wave (from [15], © Wiley Inter-Science 2005).

7.1.2 Surface waves in the grounded slab with periodic patch loading

From the previous study, we have learned that a thin grounded slab with periodic patch loading can support the propagation of surface waves. Since the substrate thickness is very thin compared to the wavelength, it is necessary to understand the properties of surface waves propagating in this structure.

First, let's examine closely the surface waves propagating in different directions. Figure 7.4 shows the dispersion curves of the first surface wave modes in the artificial ground plane. For comparison purposes, the dispersion curve of a thin grounded slab is also plotted. In the thin grounded dielectric slab, only the TM_0 mode exists at the interested frequency region (0–10 GHz). The eigen-frequency at a given propagation constant (wavenumber) is slightly lower than the light line. A similar observation is found for the periodic ground plane at a frequency below 6 GHz. As the frequency exceeds 6 GHz, the dispersion curves deviate from that of the grounded slab. Furthermore, Fig. 7.4 shows that the dispersion curve remains similar for different propagation directions, such as the x direction and the diagonal direction.

In a uniform waveguide whose cross section remains the same along the propagation direction, the surface wave can be expressed as following:

$$\vec{E}(x, y, z) = \vec{E}_0(y, z) e^{-j\beta_0 x}, \tag{7.2}$$

where x is the propagation direction and β_0 is the propagation constant. The field $\vec{E}_0(y, z)$ remains unchanged along the propagation direction. An example is the grounded dielectric slab.

For a waveguide with periodic cross section, the Floquet theory applies [13],

$$\vec{E}(x + p, y, z) = \vec{E}(x, y, z) e^{-j\beta_0 p}, \tag{7.3}$$

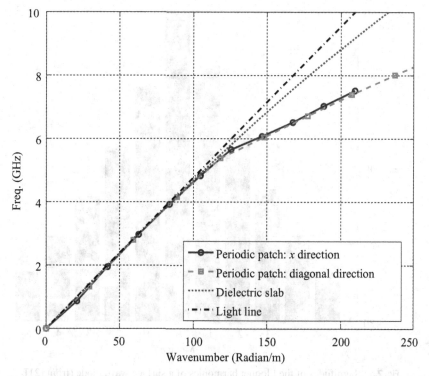

Fig. 7.4 Dispersion diagram of the thin grounded dielectric slab with periodic patch loading (from [21], © IEEE 2007).

where p is the periodicity in the waveguide. The field magnitude is no longer a constant but a periodic function in the propagation direction. The Fourier transformation is performed and the field can be expressed as the summation of an infinite series:

$$\vec{E}(x, y, z) = \sum_{n=-\infty}^{\infty} \vec{E}_n(y, z) e^{-j\beta_n x}, \quad \beta_n = \beta_0 + n\frac{2\pi}{p}, \quad (7.4)$$

$$\vec{E}_n(y, z) = \frac{1}{p} \int_0^p \vec{E}(x, y, z) e^{j\beta_n x} dx. \quad (7.5)$$

Each component $\vec{E}_n(y, z)$ is known as a Floquet harmonic and its magnitude remains the same in the propagation direction. All the Floquet harmonics must exist together to satisfy the complex periodic boundary conditions. The complete set of Floquet harmonics is referred to as one surface wave mode. Therefore, these harmonics have the same resonant frequency and their propagation constants (β_n) satisfy (7.4).

The above theory is used to process the FDTD data in order to obtain the Floquet harmonics of the surface wave propagating in the periodic ground plane. Since the surface wave is TM dominant, the E_z field is used to illustrate the wave property. Figure 7.5 shows the E_z field magnitudes for several Floquet harmonics. The propagation constant is set to 100 radian/m, which results in an eigen-frequency of 4.63 GHz. The fields are sampled at a location 0.375 mm above the bottom ground plane and normalized to the magnitude of the fundamental harmonic ($n = 0$). It is observed that the field magnitudes decrease

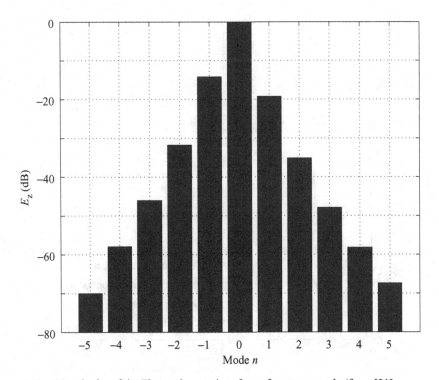

Fig. 7.5 Magnitudes of the Floquet harmonics of a surface wave mode (from [21], © IEEE 2007).

as $|n|$ increases. When $|n|$ is equal to or larger than 2, the field magnitude is 30 dB lower than the fundamental harmonic.

The E_z field distributions versus the height are calculated and plotted in Fig. 7.6. For comparison purpose, the field distribution in the thin grounded slab is also presented. The field decreases as a cosine function inside the substrate and decays as an exponential function in the air. Since the dispersion curve is close to the light line in air, the field decays slowly. A discontinuity of the E_z field is observed at the interface ($z = 3$ mm), which complies with the boundary condition. A weakly bounded surface wave exists in the thin grounded dielectric slab. The FDTD computed result agrees very well with the analytical result.

For the grounded slab loaded with the patches, the field distributions for the fundamental ($n = 0$) and two higher-order harmonics ($n = \pm 1$) are plotted in Fig. 7.6. The field of the fundamental mode has the same distribution in the dielectric as the slab waveguide, but the field in the air is noticeably weaker than the slab waveguide. For the higher-order harmonics, the fields in the substrate increase exponentially with the height z whereas the fields in the air decay exponentially with the height z. The increasing and attenuation factors in the exponential functions are determined by the propagation constant β_n. The larger the β_n is, the faster the field increases or decreases. The summation of all the harmonics satisfies the boundary conditions of the metal patches on top of the substrate. It can be concluded from Fig. 7.6 that because of the periodic patch loading,

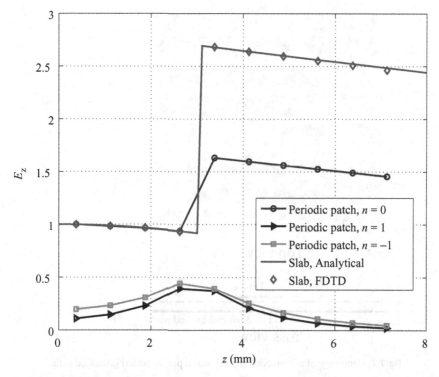

Fig. 7.6 E_z field distributions along the z direction for several surface wave modes (from [21], © IEEE 2007).

the surface wave is more tightly bounded to the artificial ground plane as compared to the traditional thin dielectric slab.

7.2 Dipole-fed surface wave antennas

In Chapter 6, the mushroom-like EBG surface is successfully used as the ground plane for a low profile dipole antenna. Upon observing the similar reflection phases but different surface wave features between the patch-loaded grounded slab and the EBG surface, it is interesting to examine the performance of a dipole near this new patch loaded grounded slab.

7.2.1 Performance of a low profile dipole on a patch-loaded grounded slab

Figure 7.7 sketches the geometry of a horizontal dipole antenna near a patch-loaded grounded slab [14–15]. To realize a low profile configuration, the horizontal dipole is positioned very close to the artificial ground plane. Note that the dipole is directly fed by a 50 Ω coaxial cable. One arm of the dipole is connected to the center conductor of the cable, and the other arm is connected to the outside conductor of the cable which is grounded to the bottom perfect electric conductor of the artificial ground plane. This feeding structure is simple to fabricate and experiments have demonstrated its applicability.

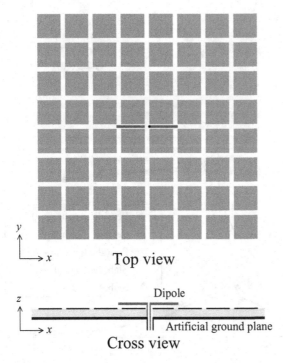

Fig. 7.7 Geometry of a horizontal dipole near a patch-loaded grounded slab.

The FDTD method is used to simulate the behavior of this radiating structure. A finite ground with a size of $2\lambda \times 2\lambda$ ($\lambda = 75$ mm) is used in the analysis, including 16×16 patches. The patch dimensions and dielectric substrate properties are the same as given in (7.1). The dipole is positioned only 0.02λ above the top surface of the artificial ground plane and the radius of the dipole is 0.005λ. Return losses of dipoles with different lengths are plotted in Fig. 7.8. The dipole length is increased from 0.06λ to 0.38λ. It is observed that when the length of the dipole is increased, the resonant frequency of the antenna decreases. The return loss value at the resonant frequency also changes with the dipole length. When the dipole length is 0.26λ, the antenna achieves a good return loss around -30 dB. It is important to point out that the length of the dipole is much smaller than the half-wavelength at the operating frequency. As a comparison, when a dipole is located near an EBG ground plane and resonates at the same frequency, the length of the dipole is 0.48λ, which is close to the half-wavelength.

To verify the simulation result, an antenna prototype is fabricated, as shown in Fig. 7.9. A RT/duroid 6002 high frequency laminate ($\varepsilon_r = 2.94 \pm 0.04$) with 120 mil (3.048 mm) thickness is used as the substrate and a 2 mm wide strip dipole is mounted 1.5 mm above the artificial ground plane. No vertical vias are used in this design.

Figure 7.10 shows the measured return loss result as compared with the simulation data. Although the dipole is very close to the ground plane, it still achieves a good return loss result (-25 dB) due to the in-phase reflection feature of the artificial ground plane. The antenna resonant frequency is 4.22 GHz, where the slight frequency shift may have

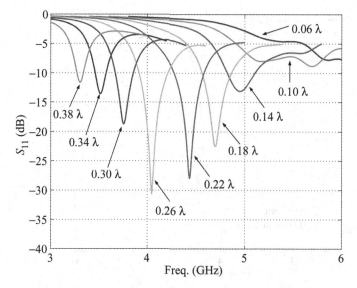

Fig. 7.8 FDTD simulated return loss results of the antenna in Fig. 7.7 with different dipole lengths (from [15], © Wiley Inter Science 2005).

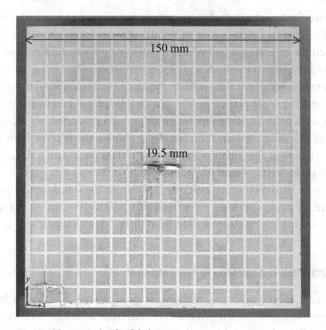

Fig. 7.9 Photograph of a fabricated antenna prototype (from [15], © Wiley Inter-Science 2005).

resulted from the numerical and fabrication errors. The bandwidth of the fabricated antenna ($S_{11} < -10$ dB) is 6.0%.

The radiation patterns of the antenna are computed and measured at its resonant frequency and the results are presented in Fig. 7.11. Several interesting observations can be made from the radiation patterns. The first observation is that the antenna shows a

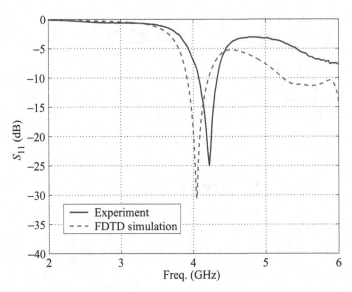

Fig. 7.10 Return loss of the antenna shown in Fig. 7.9 (from [15], © Wiley Inter-Science 2005).

low radiation power in the broadside direction ($\theta = 0°$). The main beam of this antenna points to $\theta = 50°$ direction with a gain of 4.4 dB. As a comparison, when the dipole antenna is located on an EBG ground plane, it has a maximum radiation power in the broadside direction, as shown in Fig. 6.5.

Second, E_θ is the co-polarized field in both the xz and yz plane. Similar observations are also noticed in other ϕ cut planes such as diagonal planes. It means that every φ cut plane is an E plane. In contrast, if the dipole is near an EBG ground plane, the xz plane is the E plane with E_θ as the co-polarization but the yz plane is the H plane with E_ϕ as the co-polarization. Therefore, this antenna has vastly different polarization features from a horizontal dipole on an EBG ground plane.

The xz plane pattern shows some asymmetries due to the unbalanced feed of this antenna structure. A relatively high cross polarization level in the yz plane is also observed, which is attributed to the radiation from the x-oriented dipole.

7.2.2 Radiation mechanism: the surface wave antenna

Although horizontal dipole antennas can obtain good return loss results both near the EBG and near the patch-loaded ground slab, different radiation performances are observed, such as the dipole length, beam direction, and polarization. Thus, it is important to figure out their distinct radiating mechanisms.

When a dipole is positioned on an EBG ground plane, no surface wave can be excited because of the surface wave band gap of the EBG structure. The radiation is contributed by the dipole itself. Thus, the dipole length is close to half-wavelength for resonance. The radiation of the dipole determines the broadside beam and wave polarization at different planes.

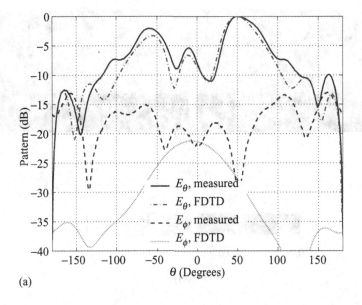

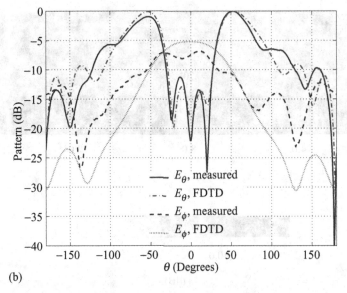

Fig. 7.11 Radiation patterns of the surface wave antenna in Fig. 7.9: (a) the xz plane pattern and (b) the yz plane pattern. The antenna gain is 4.4 dB.

In contrast, for the patch-loaded grounded slab, since the vertical vias that suppress the TM mode are removed, the band gap disappears and surface waves can propagate along the ground plane. When the coaxial-fed dipole is positioned near a patch-loaded grounded slab, strong surface waves can be excited. Actually, the unbalanced current on the vertical feeding probe activates the TM mode. Figure 7.12 shows the near field distributions of the antenna, where strong fields can be observed along the artificial ground plane. In this case, the dipole works more like a transducer rather than a radiator.

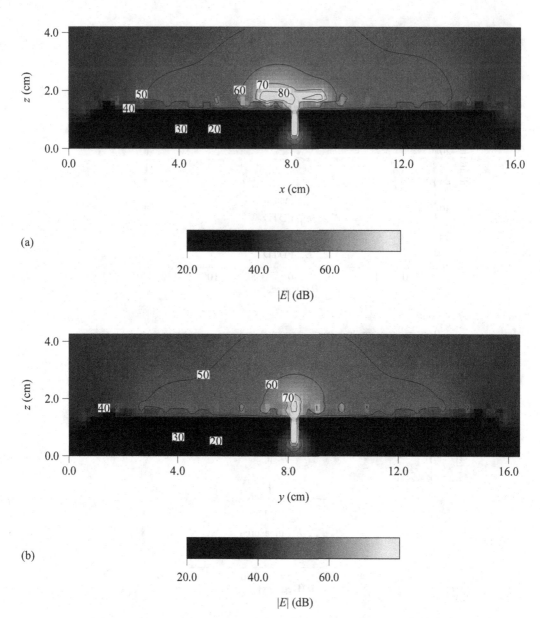

Fig. 7.12 Near field distributions of the dipole-fed surface wave antennas: (a) *xz* plane and (b) *yz* plane.

The dipole length is used to tune the impedance match of the antenna. Therefore, the length of the dipole is not necessarily half-wavelength, and the FDTD simulation shows that a proper dipole length is around 0.26 λ.

As discussed in Section 7.1, the surface waves are dominated by the TM^z modes and the electric field is vertically polarized. When the TM surface waves diffract at the boundary of the ground plane, the radiation pattern can be determined. For example, the diffractions of surface waves at the edge of the ground plane are hard diffractions, hence diffraction rays from opposite edges will cancel each other in the broadside direction, resulting in a radiation null. The vertically polarized surface waves also determine that the diffraction field must be polarized along the θ direction. Thus, a monopole-like radiation is generated, as observed in Fig. 7.11.

In summary, the diffraction of the surface waves is the main contributor for the antenna radiation. Therefore, this antenna structure can be identified as a surface wave antenna (SWA). Compared to traditional surface wave antenna structures, this antenna design uses the novel artificial ground plane so that an attractive low profile configuration is achieved. This feature is desirable in modern wireless communication systems.

7.2.3 Effect of the finite artificial ground plane

To further understand the radiation mechanism of the surface wave antennas, the ground plane effect is investigated here. Note that the diffraction of surface waves at the boundary of the ground plane plays an important role in determining the antenna patterns. Radiation performances of three antennas with different ground planes are calculated and compared, as shown in Fig. 7.13:

Case 1: $2\lambda \times 2\lambda$ ground plane including 16×16 patches;
Case 2: $2\lambda \times 2\lambda$ ground plane including 8×8 patches in the center;
Case 3: $1\lambda \times 1\lambda$ ground plane including 8×8 patches.

The dipole dimensions, patch size, and dielectric substrate properties all remain the same. It is observed that the reduction of either patch numbers or ground plane size has little effect on the local excitation of the surface waves, thus all three antenna cases have very similar resonant frequencies and return loss results.

Figure 7.13b compares the radiation patterns of these three surface wave antennas in the yz plane. All antennas exhibit monopole-like radiation patterns. A low radiation power appears in the broadside direction and E_θ is the co-polarized field. Similar radiation patterns are observed in other ϕ cut planes.

The radiation patterns of cases 1 and 2 are very similar to each other, such as the beam direction and the back lobes. This indicates that the periodic square patches are helpful for the surface wave excitation from the dipole. Once the surface waves are excited, they can propagate along the grounded slab without patches due to the similar dispersion curves in Fig. 7.4. Therefore, one only needs to design periodic patches locally underneath the dipole. Further FDTD simulations show that a minimum of three rows of patches (6×6) are required to maintain the same radiation performance.

When the ground plane size is increased as from case 3 to case 1, the radiation pattern varies and the beam direction moves toward the low elevation angle. The antenna beams of the $1\lambda \times 1\lambda$ ground plane case point to the $\theta = 35°$ direction whereas the antenna beam of the $2\lambda \times 2\lambda$ ground plane case points to the $\theta = 50°$ direction. FDTD simulation of a surface wave antenna with a $3\lambda \times 3\lambda$ ground plane is also performed, and the beam points to the $\theta = 60°$ direction. Thus, the beam direction is determined by the ground plane size. The beam direction change of the SWA is similar to a vertical monopole on a

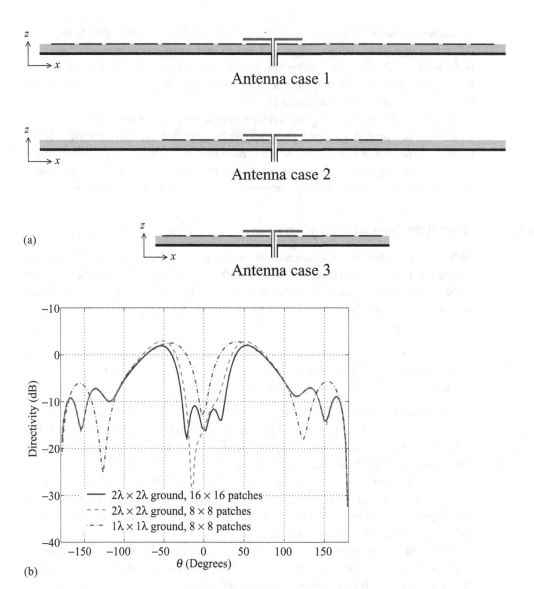

Fig. 7.13 Surface wave antenna with different ground planes. (a) Geometries: case 1 with $2\lambda \times 2\lambda$ ground plane size and 16×16 patches, case 2 with $2\lambda \times 2\lambda$ ground plane size and 8×8 central patches, and case 3 with $1\lambda \times 1\lambda$ ground plane size and 8×8 patches. (b) Radiation patterns of the SWAs with different ground planes.

finite ground plane. When an infinite ground plane is used, the beam direction will go to $\theta = 90°$, the same as a vertical monopole on an infinite PEC ground plane.

It is worthwhile to point out that the above results on the ground plane size can be used as a reference for surface wave antenna applications in practical environments. For example, when a surface wave antenna is mounted on the roof of a car, the beam direction is not determined by the size of the surface wave antenna itself, but by the overall dimensions of the conducting ground plane: the car roof.

7.2.4 Comparison between the surface wave antenna and vertical monopole antenna

As we have demonstrated, the radiation pattern of this surface wave antenna is similar to a vertical monopole antenna. Thus, it is instructive to compare the surface wave antenna and an actual vertical monopole antenna to appreciate their resemblance. To this end, a vertical monopole on a finite PEC ground plane is designed and its radiation characteristics are calculated for comparison. The ground plane has the same size of $2\lambda \times 2\lambda$ and the radius of the monopole is 0.005λ. The monopole length is tuned to 0.22λ so that it has a similar resonant frequency to the surface wave antenna, as shown in Fig. 7.14a. The monopole antenna has a bandwidth of 20.6%, which is wider than the surface wave antenna.

Fig. 7.14b compares the radiation patterns of both antennas in the yz plane because E_θ polarization is of particular interest in this plane. It is noticed that the two antennas have similar shapes of radiation patterns. For example, the vertical monopole has the same beam direction at $\theta = 50°$ as the surface wave antenna. Both the monopole and surface wave antennas have E_θ as the co-polarized fields. The back lobes of the two antennas are also close to each other: both the field levels and null positions. Compared to the monopole antenna, the surface wave antenna has a relatively higher cross-polarization level and a slightly lower directivity, which is due to the direct radiation from the dipole. These deficiencies can be improved using a symmetric feeding structure such as a circular disk to replace the dipole.

The attractive feature of the surface wave antenna is the low profile configuration. The height of the vertical monopole antenna is 0.22λ whereas the height of the horizontal dipole over the artificial ground plane is only 0.02λ. Thus, the dipole height is less than 10% of the monopole antenna. Even if the thickness of the substrate is considered, the overall antenna height is 0.06λ, which is still much smaller than the monopole. Therefore, the low profile surface wave antenna with a monopole-like radiation pattern has a promising potential in wireless communications such as satellite radio systems for vehicles.

7.3 Patch-fed surface wave antennas

The linearly polarized and omni-directional radiation pattern of a vertical monopole antenna is widely used in wireless communications such as radio broadcast and wireless local area network (WLAN). Since the vertical monopole antenna is not recommended when low profile or conformal geometry is desired, circular patch and annular-ring microstrip antennas operating at higher-order modes are designed to realize the monopole-like pattern [16–17]. However, these microstrip antenna designs usually suffer from a narrow impedance bandwidth because of the high Q resonant feature.

In the preceding section, a dipole-fed surface wave antenna is proposed as an alternative approach to generate a monopole-like radiation pattern with a low profile configuration. However, the cross polarization in this design is relatively high due to the dipole feed, as shown in Fig. 7.14. To solve this problem, a surface wave antenna excited by a circular patch is presented here. It radiates a similar pattern to a vertical monopole antenna with

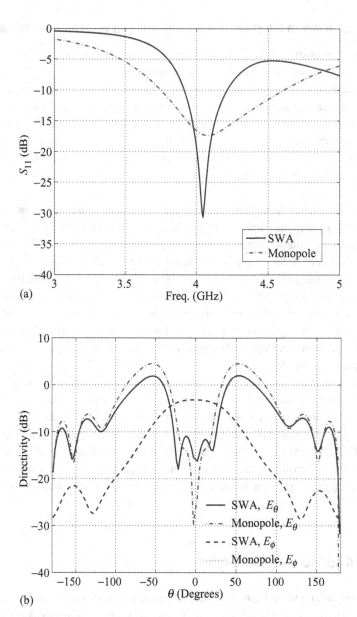

Fig. 7.14 Comparisons between the surface wave antenna and a vertical monopole antenna: (a) return losses and (b) radiation patterns.

a low cross polarization and a good impedance bandwidth. The antenna performance is numerically analyzed and experimentally verified.

7.3.1 Comparison between a circular microstrip antenna and a patch-fed SWA

The geometry of the proposed patch-fed surface wave antenna (PFSWA) is shown in Fig. 7.15 [18]. The PFSWA consists of two parts: an artificial ground plane and a circular excitation patch. The artificial ground plane is a thin grounded dielectric slab loaded with

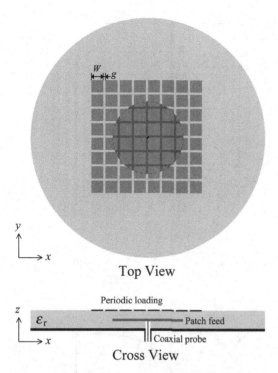

Fig. 7.15 Geometry of a low profile patch-fed surface wave antenna (PFSWA) that realizes a monopole-like radiation pattern (from [18], © IET 2007).

periodic square patches. The dimensions of the artificial ground plane are the same as (7.1). As revealed before, strong surface waves can be excited and propagate within this ground plane. A circular patch is located in the middle of the substrate to excite the surface waves. It has a radius of 21 mm and height of 1.5 mm. A 50 Ω coaxial probe is connected to the center of the circular patch for a symmetric excitation. In order to get a symmetric diffraction pattern, the ground plane is truncated into a circular shape. The radius of the ground plane is 75 mm.

The FDTD simulated input impedance and return loss of the PFSWA are presented in Fig. 7.16. For the purpose of comparison, a conventional center-fed circular patch antenna, which also radiates a monopole-like pattern, is simulated as well. The microstrip antenna has the same substrate properties and circular patch parameters as the PFSWA, but without periodic square patch loading.

It is observed that the microstrip antenna has a high input impedance and a high Q factor, resulting in a poor return loss of only −6 dB. The reason is that the TM_{02} mode of the circular patch has a large E field but a small H field in the center of the patch. Thus, inherently high impedance is obtained. Although one can move the feed location to reduce the impedance, it will increase the cross polarization in the radiation pattern.

In contrast, the PFSWA resonates at 4.72 GHz with a good return loss near −30 dB. Both the input impedance and Q factor of the PFSWA are smaller than the microstrip antenna. These improvements are obtained because of the efficient launching of the surface waves in the artificial ground plane. The periodic square patches successfully

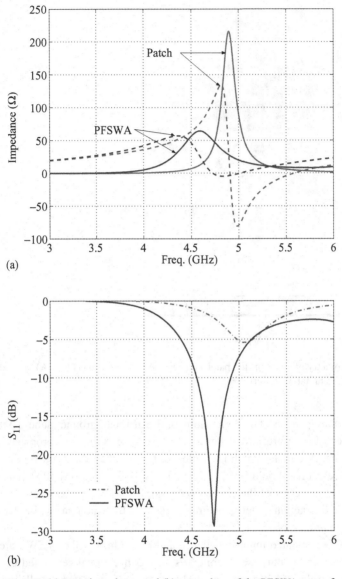

Fig. 7.16 (a) Input impedance and (b) return loss of the PFSWA to a reference center fed circular patch antenna (from [18], © IET 2007).

convert the electromagnetic fields underneath the circular patch into the surface waves propagating along the ground plane. Thus, the Q factor of the entire radiating structure is reduced and a better return loss is obtained.

To further understand the different radiation mechanisms between the microstrip antenna and the surface wave antenna, the near field distributions of both antennas are calculated at the resonant frequency and graphically presented in Fig. 7.17. It is observed that the energy of the microstrip antenna is almost confined underneath the circular patch ($r = 2.1$ cm). The field outside the patch is low and the value at the edge of the ground plane ($R = 7.5$ cm) is around -40 dB. As a comparison, when the periodic

7.3 Patch-fed surface wave antennas

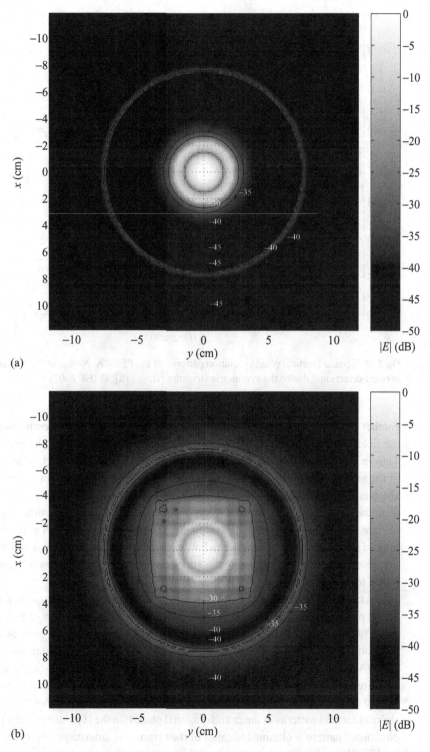

(a)

(b)

Fig. 7.17 Near field distributions of (a) the patch antenna and (b) the PFSWA. It is observed that stronger surface waves are excited in the PFSWA (from [18], © IET 2007).

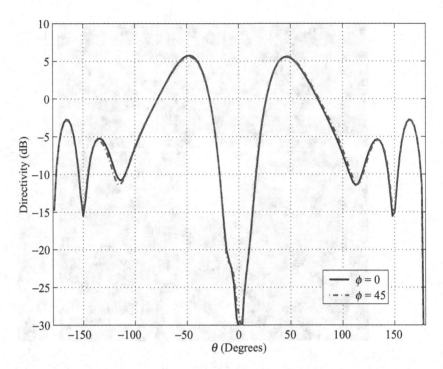

Fig. 7.18 E plane (vertical plane) radiation patterns of the PFSWA. Note that the H plane pattern is omni-directional due to the symmetric structure (from [18], © IET 2007).

patches are loaded on top of the circular patch, energy can be efficiently coupled from the field underneath the circular patch to the surface waves outside the circular patch. Thus, a stronger electric field is observed in Fig. 7.17b and the field value at the edge of the ground plane is around −35 dB, which is 5 dB higher than that of the microstrip antenna. Since both antennas are normalized to the same delivered power, it is clear from the comparison that the PFSWA is more efficient to excite the surface waves. The surface waves propagate along the artificial ground plane and diffract at the edge of the ground plane, resulting in a higher radiation efficiency. The comparison of the near field distributions clearly explains why the surface wave antenna has a lower Q and a better return loss.

The radiation pattern of the PFSWA is computed at the resonant frequency 4.72 GHz. The E plane patterns are plotted in Fig. 7.18. Both the xz plane ($\phi = 0°$) and diagonal plane ($\phi = 45°$) patterns are presented, and the yz plane pattern is omitted because it is identical to the xz plane pattern due to the symmetry of the antenna geometry. The PFSWA has a deep null in the broadside direction and the antenna beam is at the $\theta = 46°$ direction. The E-field is polarized along the θ direction and the cross polarization is −40 dB lower than the co-polarization level. It is also observed that the antenna has almost identical patterns in the xz and diagonal planes. In the H plane (xy plane), an omni-directional pattern is obtained because of the symmetric antenna geometry. In summary, the PFSWA realizes a monopole-like radiation pattern.

It is worthwhile to emphasize that the thickness of the PFSWA is only 3 mm, which is less than 0.05 λ. Thus, it is 80% shorter than a typical quarter-wavelength monopole antenna. The low profile configuration of the PFSWA is desirable for many wireless communication environments where a conformal geometry is required, such as automobiles and vessels.

7.3.2 Experimental demonstration

To demonstrate the concept of the patch-fed surface wave antenna, an antenna prototype that has the same dimensions as the numerical study is fabricated and tested. Figure 7.19a shows the photograph of the excitation of the circular patch fabricated on a 60 mil (1.524 mm) thick RT/duroid 6002 high frequency laminate ($\varepsilon_r = 2.94 \pm 0.04$). A 50 Ω SMA connector is soldered to the center of the circular patch. Figure 7.19b is a photograph of the 8 × 8 periodic patches loading fabricated on another dielectric slab. These two layers are stacked together to form a complete PFSWA structure, as shown in Fig. 7.15.

Figure 7.20a presents the measured return loss of the PFSWA compared to the FDTD simulation result. According to the measured result, the antenna resonates at 4.74 GHz with a good return loss of −28 dB, which agrees well with the FDTD simulation result. The impedance bandwidth of the fabricated antenna ($S_{11}<-10$ dB) is 5.6%.

The radiation patterns of the PFSWA are measured at the resonant frequency of 4.74 GHz and plotted in Fig. 7.20b. As expected from the FDTD simulation, a monopole-like pattern is obtained. Both the xz plane ($\phi = 0°$) and diagonal plane ($\phi = 45°$) patterns are presented, and they are almost identical to each other. The patterns have a deep null in the broadside direction and the antenna beam is along the $\theta = 47°$ direction with a gain of 5.6 dB. The co-polarization is along the θ direction, and the cross polarization is −25 dB lower than the co-polarization in the broadside. A noticeable cross polarization is observed in the back side of the antenna due to scattering from the supporting posts and feeding cable. The presented experimental results verify the numerical results and prove the concept and radiation performance of the low profile patch-fed surface wave antenna.

7.4 Dual band surface wave antenna

In Section 7.2.4, we compared the performance of a dipole-fed surface wave antenna with a vertical monopole antenna. It was noticed that the SWA realizes a monopole-like radiation pattern with an attractive low profile configuration. However, there are two deficiencies with the dipole-fed SWA, namely, the relatively high cross polarization and the narrow bandwidth. In the previous section, a patch-fed SWA is proposed, which successfully suppresses as the cross polarization. To solve the second problem, wideband or multi-band techniques [19–20] needs to be developed to increase the effectiveness of the surface wave antennas in wireless communication systems.

In this section, a novel dual band surface wave antenna (DBSWA) is proposed. The DBSWA is excited by a center fed cross strip with split branches. The length of the split

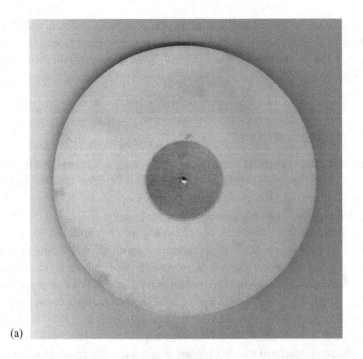

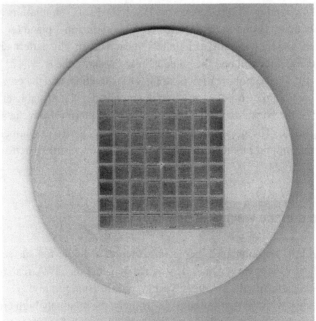

Fig. 7.19 Photographs of a fabricated PFSWA prototype: (a) a center-fed circular patch in the middle layer, and (b) periodic square patch array on the top layer. The overall antenna thickness is 3 mm (from [18], © IET 2007).

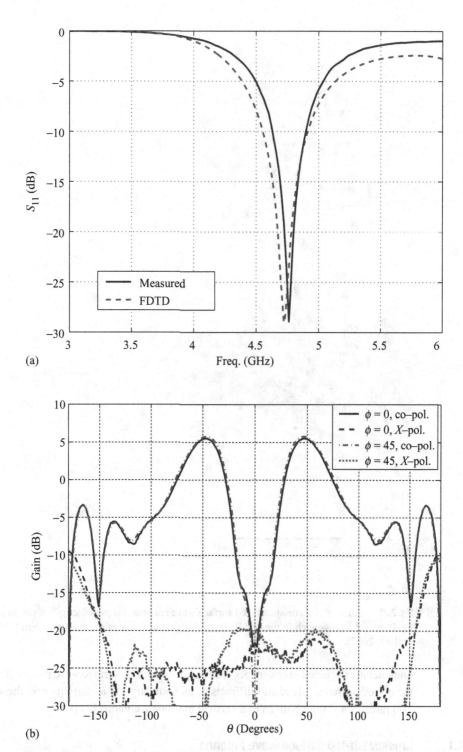

Fig. 7.20 Measured results of the fabricated PFSWA prototype: (a) return loss and (b) radiation patterns (from [18], © IET 2007).

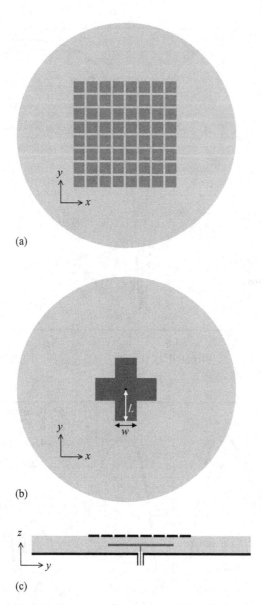

Fig. 7.21 Geometry of a crosspatch-fed surface wave antenna: (a) periodic square patches on the top layer, (b) a crosspatch in the middle layer, and (c) side view of the antenna (from [21], © IEEE 2007).

branches are designed to cover specified frequencies. The surface wave propagates along the periodic ground plane and diffracts at its boundary. As a consequence, the antenna also radiates a similar pattern to a vertical monopole antenna [21]).

7.4.1 Crosspatch-fed surface wave antenna

First, let's investigate a crosspatch-fed SWA [21], as shown in Fig. 7.21. 8×8 square patches are printed on top of the substrate. To excite the surface waves, a crosspatch with

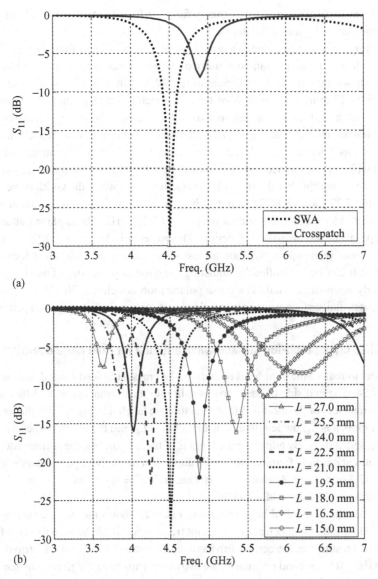

Fig. 7.22 Simulated return loss of the surface wave antenna: (a) comparison between an SWA and a crosspatch antenna and (b) SWA with different cross dipole lengths (from [21], © IEEE 2007).

a center probe is inserted in the middle of the substrate. The length L of the crosspatch is 21 mm and the width W is 15 mm. The ground plane is truncated into a circular shape with a radius of 75 mm. The thickness of the entire surface wave antenna is 3 mm.

The radiation performance of the SWA is simulated using the Ansoft HFSS commercial software [22]. The computed return losses of the SWA and a crosspatch antenna (without periodic patch loading) are compared in Fig. 7.22a. The reflection coefficient for the crosspatch antenna is only −8 dB, while for the SWA is −28 dB. Again, this return loss improvement is due to the efficient launching of the surface wave in the artificial ground plane. The surface wave receives energy from the feeding structure, resulting in a better reflection coefficient of the antenna.

An important design issue is how to control the operational frequency of the surface wave antenna. Thus, a parametric study of the feed structure is performed here. Figure 7.22b shows the reflection coefficient of the SWA with different length L varying from 15 mm to 27 mm. The width W of the crosspatch is set to 15 mm. It is observed that as the length L of the crosspatch increases, the resonant frequency decreases. The best matching occurs when the length of the crosspatch is 21 mm, and the reflection coefficient increases as L increases or decreases referring to this length. Parametric study on the patch width is also carried out. It is noticed that the resonant frequency remains similar for different widths, but the return loss value improves when the width is increased.

Figure 7.23 shows HFSS computed directivity of this surface wave antenna ($L = 21$ mm, $w = 15$ mm) at the resonant frequency of 4.5 GHz. The E plane patterns exhibit a deep null in the broadside direction. The patterns in the xz plane ($\phi = 0°$) and in the diagonal plane ($\phi = 45°$) are almost identical. The main beam is located around 45° and it can be controlled by adjusting the ground plane size. The H plane pattern is nearly omni-directional. The cross polarization is below -20 dB. In summary, the crosspatch-fed surface wave antenna also has a monopole-like radiation pattern.

7.4.2 Modified crosspatch-fed surface wave antenna for dual band operation

In order to realize the dual band operation, the crosspatch feed is modified, as shown in Fig. 7.24. The end side of each strip is split into three branches with different lengths. The parameters of the modified feed structure are labeled in Fig. 7.24b: the length of the center branch is L_1, the length of side branches is L_2, and the width of the strip branches are w_1, w_2, and w_3. By adjusting the lengths of these branches, the surface wave antenna resonates at multiple frequencies. An example design for these parameters is as follows: $L_0 = 13.5$ mm, $L_1 = 22$ mm, $L_2 = 20$ mm, and $w_1 = w_2 = w_3 = 3$ mm. This antenna resonates at 4.15 GHz and 4.79 GHz.

To understand the dual band operation, Fig. 7.25 shows the surface current densities on the modified crosspatch at two resonant frequencies. It can be seen from the figure that at the first resonance, the center strip has a high current density, and this strip resonates at 4.15 GHz. At the second resonance, the two outer strips have a high current density. Since their length is shorter than the center strip, the antenna resonates at a higher frequency of 4.79 GHz. It's clear from these current plots that the modified patch resonates at dual frequency bands because of the different strip lengths.

An antenna prototype is fabricated and tested to demonstrate the concept of this dual band surface wave antenna. Figure 7.26a shows the photograph of the modified crosspatch fabricated on a 1.524 mm thick RT/duroid 6002 high frequency laminate ($\varepsilon_r = 2.94$). A 50 Ω SMA connector is soldered to the center of the patch. Figure 7.26b is a photograph of the 8 × 8 periodic patch loading fabricated on another dielectric slab. These two layers are stacked together to form a complete dual band surface wave antenna (DBSWA).

For the purpose of comparison, a conventional center-fed modified crosspatch microstrip antenna, which also radiates a monopole-like pattern, is fabricated and

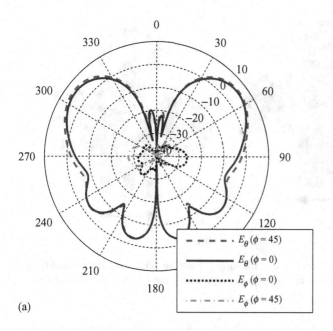

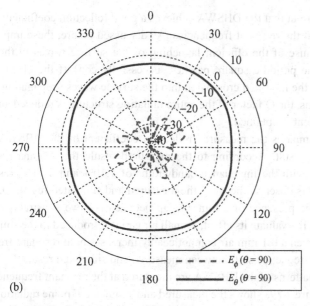

Fig. 7.23 Radiation patterns of the surface wave antenna: (a) E plane (vertical plane) patterns and (b) H plane (*xy* plane) patterns (from [21], © IEEE 2007).

measured as well. The microstrip antenna has the same substrate properties and patch dimensions as the DBSWA, but without periodic square patch loading. It is observed that the microstrip antenna resonates at 4.7 and 5.4 GHz with poor return loss, as shown in Fig. 7.27. In contrast, the fabricated DBSWA resonates at 4.2 and 4.9 GHz. These resonant frequencies are lower than the resonant frequencies of the microstrip antenna. It is

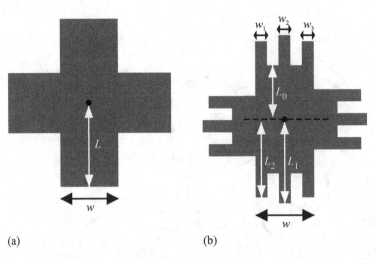

Fig. 7.24 Dual band operation of the surface wave antenna is realized by modifying the feed structure from (a) a simple crosspatch to (b) a crosspatch with split branches (from [21], © IEEE 2007).

important to point out that the DBSWA achieves a good reflection coefficient near −20 dB and −25 dB at the resonant frequencies. As discussed before, these improvements are obtained because of the efficient launching of the surface waves in the artificial ground plane. The periodic square patches successfully convert the electromagnetic fields underneath the modified crosspatch into the surface waves propagating along the ground plane. Thus, the Q factor of the entire radiating structure is reduced and a better reflection coefficient is obtained.

Figure 7.28 compares the measured reflection coefficient of the DBSWA with the HFSS simulation result. According to the measured result, the antenna resonates at 4.2 and 4.9 GHz with the impedance bandwidths of 1.10% and 3.94%, respectively. A frequency shift is observed between the measured and simulated results. The reason for this shift is the presence of a thin air gap between the two substrates during the stacking process. To evaluate its effect, a small air gap is introduced in the simulation. It is observed that even a 0.1 mm air gap noticeably increases both resonant frequencies. This explains the frequency shift between measured and simulated results.

The radiation patterns of the DBSWA are measured at the resonant frequencies of 4.2 and 4.9 GHz. Figure 7.29a shows the measured and simulated E plane radiation patterns at 4.2 GHz. It has a deep null in the broadside direction and the main beam is at 55° with a gain of 4.9 dB. The co-polarization is along the θ direction, and the cross polarization is 20 dB lower than the co-polarization. The measured and simulated radiation patterns at 4.9 GHz are shown in Fig. 7.29b. Also, it has a deep null in the broadside direction and the main beam is at 45° with a gain of 5.4 dB. The increment in the antenna gain is due to the fact that the ground plane is relatively larger in terms of wavelength at the higher resonant frequency. The H plane pattern is nearly omni-directional, as shown in Fig. 7.30. The cross polarization level is more than 15 dB below that of the co-polarization.

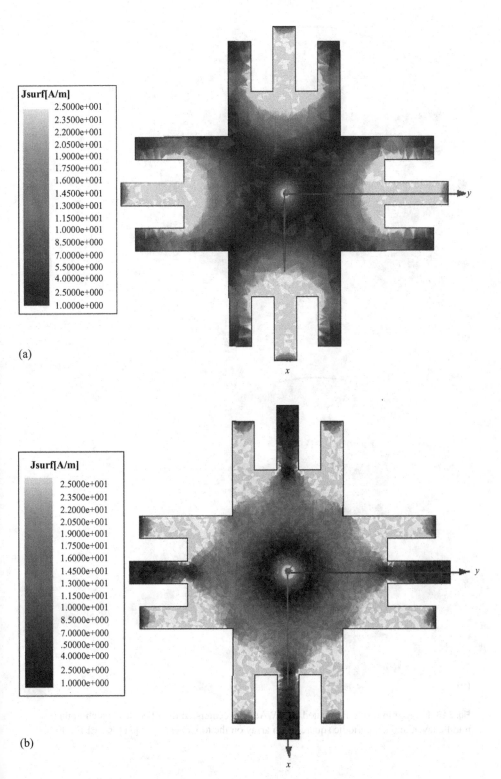

Fig. 7.25 Surface current densities on the modified crosspatch SWA at (a) 4.15 GHz and (b) 4.79 GHz (from [21], © IEEE 2007).

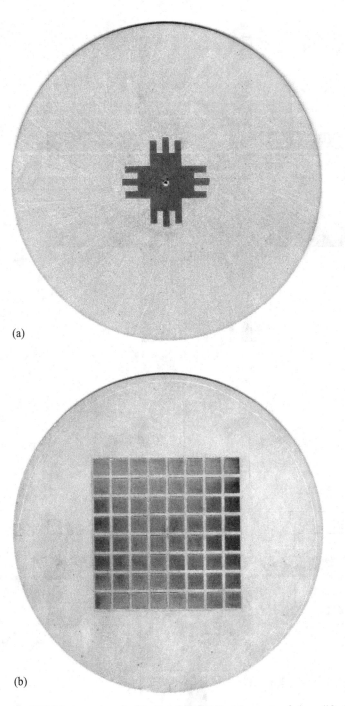

Fig. 7.26 Photographs of a fabricated DBSWA: (a) a center-fed modified crosspatch in the middle layer, and (b) periodic square patch array on the top layer (from [21], © IEEE 2007).

7.4 Dual band surface wave antenna

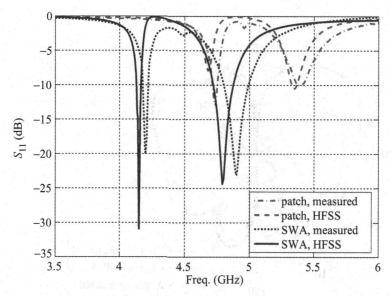

Fig. 7.27 Return loss comparison between the dual band surface wave antenna and the patch antenna (from [21], © IEEE 2007).

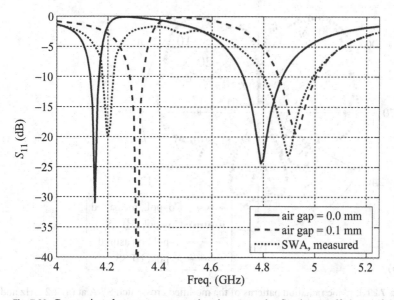

Fig. 7.28 Comparison between computed and measured reflection coefficients of the dual band SWA (from [21], © IEEE 2007).

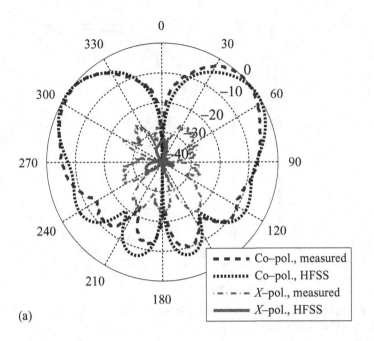

(a)

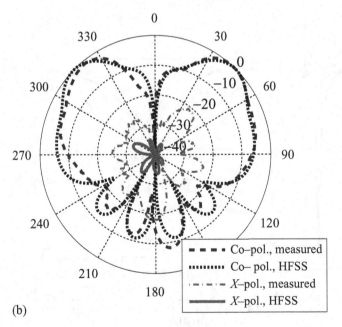

(b)

Fig. 7.29 E plane radiation patterns of the modified crosspatch SWA at (a) 4.2 GHz and (b) 4.9 GHz (from [21], © IEEE 2007).

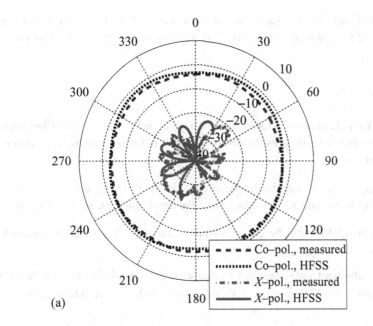

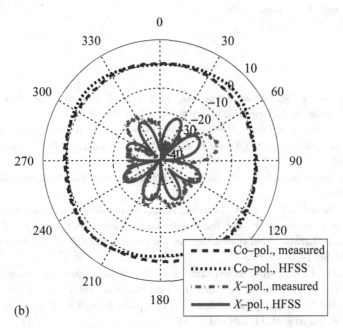

Fig. 7.30 H plane radiation patterns of the modified crosspatch SWA at (a) 4.2 GHz and (b) 4.9 GHz (from [21], © IEEE 2007).

In summary, the measured radiation patterns agree well with the HFSS simulations, and a monopole-like pattern is obtained at both frequencies of this surface wave antenna.

7.5 Projects

Project 7.1 A dipole antenna is located above a grounded dielectric slab. Characterize the antenna performance such as the return loss and radiation patterns with different feeding mechanisms:

(a) Using a gap source to feed the dipole antenna, as shown in Fig. 6.1.
(b) Using an un-balanced coaxial probe to feed the dipole antenna, as shown in Fig. 7.7.

The effects of dipole length, height, substrate thickness and dielectric constant should be considered in the project.

Project 7.2 Multi-band and broadband surface wave antenna (SWA) designs. Based on the SWAs shown in Fig. 7.15 and Fig. 7.17, design new surface wave antenna geometries with

(a) a multiple resonant behavior, such as tri-band operation;
(b) a broad impedance bandwidth.

References

1. Francis J. Zucker, "Surface-wave antennas," in *Antenna Engineering Handbook*, 3rd edn., Richard C. Johnson, McGraw-Hill Inc., 1993.
2. F. Schwering and A. A. Oliner, "Millimeter-wave antennas," in *Antenna Handbook, Theory, Applications, and Design*, Y. T. Lo and S. W. Lee, Van Nostrand Reinhold Company Inc., New York, 1988.
3. R. Elliot, "Spherical surface wave antennas," *IRE Trans. Antennas Propagat.*, **vol. 4, no. 3**, 422–8, 1956.
4. R. Hougardy and R. C. Hansen, "Scanning surface wave antennas – oblique surface waves over a corrugated conductor," *IRE Trans. Antennas Propagat.*, **vol. 6, no. 4**, 370–6, 1958.
5. L. B. Felson, "Radiation from a tapered surface wave antenna," *IRE Trans. Antennas Propagat.*, **vol. 8**, 577–86, 1960.
6. F. J. Zucker and J. A. Storm, "Experimental resolution of surface wave antenna radiation into feed and terminal patterns," *IEEE Trans. Antennas Propogat.*, **vol. 18**, 420–2, 1970.
7. C.-L. Chi and N. G. Alexopoulos, "Radiation by a probe through a substrate," *IEEE Trans. Antennas Propagat.*, **vol. 34**, 1080–91, 1993.
8. G. Fikioris, R. W. P. King, and T. T. Wu, "Novel surface wave antennas," *IEE Proc. Microw. Antennas, Propagat.*, **vol. 143, no. 1**, 1–6, 1996.
9. J. P. Kim, C. W. Lee, and H. Son, "Analysis of corrugated surface wave antenna using hybrid MOM/UTD technique," *Electronic Lett.*, **vol. 35, no. 5**, 353–4, 1999.
10. T. Zhao, D. R. Jackson, J. T. Williams, and A. A. Oliner, "General formulas for 2-D leaky-wave antennas," *IEEE Trans. Antennas Propogat.*, **vol. 53, no. 11**, 3525–33, 2005.

11. A. Aminian, F. Yang, and Y. Rahmat-Samii, "Bandwidth determination for soft and hard ground planes by spectral FDTD: a unified approach in visible and surface wave regions," *IEEE Trans. Antennas Propagat.*, **vol. 53, no. 1**, 18–28, 2005.
12. F. Yang, J. Chen, Q. Rui, and A. Elsherbeni, "A simple and efficient FDTD/PBC algorithm for periodic structure analysis," *Radio Sci.*, **vol. 42, no. 4**, RS4004, 2007.
13. K. Zhang and D. Li, *Electromagnetic Theory for Microwaves and Optoelectronics*, 2nd edn., Publishing House of Electronics Industry, 2001.
14. F. Yang, A. Aminian, and Y. Rahmat-Samii, "A low profile surface wave antenna equivalent to a vertical monopole antenna," *2004 IEEE APS Int. Symp. Dig.*, vol. 2, pp. 1939–42, Monterey, CA, June 20–26, 2004.
15. F. Yang, A. Aminian, and Y. Rahmat-Samii, "A novel surface wave antenna design using a thin periodically loaded ground plane," *Microwave Optical Tech. Lett.*, **vol. 47, no. 3**, 240–5, 2005.
16. J. Huang, "Circularly polarized conical patterns from circular microstrip antennas," *IEEE Trans. Antennas Propagat.*, **vol. 32**, 991–4, 1984.
17. L. Economou and R. J. Langley, "Patch antenna equivalent to simple monopole," *Electronic Lett.*, **vol. 33, no. 9**, 727–9, 1999.
18. F. Yang, Y. Rahmat-Samii, and A. Kishk, "A low profile surface wave antenna for wireless communications," *IET Proceedings Microwaves Antennas & Propagation*, vol. 1, no. 1, pp. 261–6, February 2007.
19. F. Yang, A. Al-Zoubi, and A. Kishk, "A dual band surface wave antenna with a monopole like pattern," *2006 IEEE APS Int. Symp. Dig.*, vol. 5, pp. 4281–4, July 2006.
20. C. B. Ravipati and C. J. Reddy, "Dual-band planar antennas with monopole radiation patterns," *2006 IEEE APS Int. Symp.*, pp. 469, July 2006.
21. A. Al-Zoubi, F. Yang, and A. Kishk, "A low profile dual band surface wave antenna with a monopole like pattern," *IEEE Trans. Antennas Propagat.*, **vol. 55, no.12**, 3404–12, December 2007.
22. *HFSS: High Frequency Structure Simulator Based on Finite Element Method*, 2004, v. 9.2.1, Ansoft Corporation.

Appendix EBG literature review

1 Overview

In recent years, electromagnetic band gap (EBG) structures have attracted increasing interest in the electromagnetic community. Because of their desirable electromagnetic properties, they have been widely studied for potential applications in antenna engineering. Hundreds of EBG papers have been published in various journals and conferences. To illustrate the rapid increase of research interest, a simple search using the keywords "EBG" and "antenna" was performed on IEEE Xplore on 1/18/2008 and the data are plotted in Fig. A.1 based on the available number of publications. It is clear that the number of papers has steadily increased over the years with an apparent peak of publications in 2005.

Here we provide a comprehensive overview of these publications so that readers can establish a clear picture of EBG development. It will also help readers to easily find papers related to their own research interests. Since the scope of this book is EBG metamaterials, papers on double negative property, left-handed propagation, and negative refractive index are not covered in this overview. We also regret if we have missed some papers as there have been so many international conferences with sessions on this topic.

To start with, we would like to emphasize several special issues organized in microwave and antenna journals that relate to EBG research. These special issues provide the readers with background and information for EBG research. They summarize the latest research progress at the time of publication, and have greatly stimulated EBG research afterwards. In particular, the editorial comments are interesting and worthwhile to read. A list of these special issues is provided below:

(1) *IEEE Trans. Microwave Theory Tech., Special Issue on Electromagnetic Crystal Structures, Designs, Synthesis, and Applications*, edited by Axel Scherer, Theodore Doll, Eli Yablonovitch, Henry Everitt, and Aiden Higgins, vol. 47, no. 11, November 1999.
(2) *IEEE Trans. Antennas Propagat., Special Issue on Meta-materials*, vol. 51, no. 10, edited by Richard Ziolkowski and Nader Engheta, October 2003.
(3) *IEEE Trans. Antennas Propagat., Special Issue on Artificial Magnetic Conductors, Soft/Hard Surfaces and other Complex Surfaces*, edited by Per-Simon Kildal, Ahmed Kishk, and Stefano Maci, vol. 53, no. 1, January 2005.
(4) *IEEE Trans. Microwave Theory Tech., Special Issue on Metamaterial Structures, Phenomena, and Applications*, edited by Tatsuo Itoh and Arthur A. Oliner, vol. 53, no. 4, April 2005.

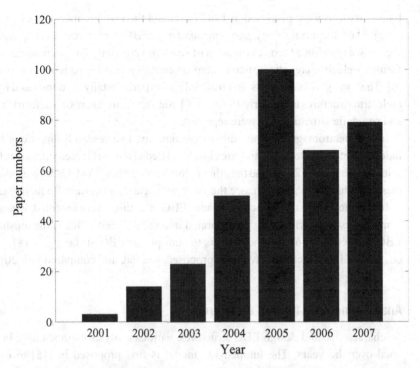

Fig. A.1 Number of EBG-antenna papers published per year. Data obtained from IEEE Xplore on 1/18/2008.

(5) *IET Microwaves Antennas Propagat., Special Issue on Metamaterials*, edited by George Eleftheriades and Yiannis Vardaxoglou, vol. 1, no. 1, January 2007.

In 2006, *Metamaterials: Physics and Engineering Explorations*, edited by Nader Engheta and Richard Ziolkowski, was published by IEEE-Wiley. The book explores the fundamental physics and the engineering designs of metamaterials, and points to a wealth of state-of-the-art applications for antennas, waveguides, devices, and components. Half of this book is focused on EBG characterizations and applications.

In the following sections, we categorize the EBG papers into three major groups: (1) EBG theories, (2) EBG designs, and (3) EBG applications in antenna engineering. In each group, papers on similar topics are put together and sorted chronologically. Although we have attempted to include as complete references as possible, it is inevitable that we might have missed some interesting papers due to the fast progress of research in the EBG area. We would appreciate readers' help to inform us of any missing references.

2 EBG theories

2.1 Origins of EBG

The concept of electromagnetic band gap (EBG) structures originates from the solid-state physics and optic domain, where photonic crystals with forbidden band gap for light emissions were proposed in 1987 [1–2] and then widely investigated in the 1990s [3–5]. Thus, the terminology, *photonic band gap (PBG)* structures, was popularly used in the early days. Since then, a profusion of scientific creativity has been witnessed as new forms of electromagnetic structures are invented for radio frequency and microwaves [6]. The band gap feature was first realized and experimentally demonstrated by periodic dielectric structures in the early 1990s [7–8]. Subsequently, arrays of dielectric rod [9–10] and woodpile structure [11] were reported.

In the electromagnetic and antenna community, two research directions have been independently developed, which are closely related to the EBG research. One is frequency selective surfaces [12–13] and the other is soft/hard surfaces [14]. Using periodic designs, these structures can stop/enhance the wave propagation at certain frequency ranges.

In the late 1990s, two important planar EBG structures were invented, where metallic components were effectively incorporated into the unit cells. One is the mushroom-like EBG surface [15–16] and the other is the uni-planar EBG surface [17–18]. From then on, many EBG structures have been proposed, studied, and compared [19–20].

2.2 Analysis methods for EBG structures

To characterize and design EBG structures, various analysis methods have been developed over the years. The lumped LC model is first proposed in [15] to explain the working mechanism of EBG structures and to predict their operational bands. Following that, several simple and accurate circuit models are presented [21–22]. Besides the circuit models, EBG structures are also analyzed using transmission line models [23–24] and the resonant cavity model [25]. A network theory is presented in [26] to analyze waves propagating along arbitrary directions on EBG surfaces while an artificial neural network (ANN) theory is developed in [27] for EBG analysis. Plane wave expansion method [28] and spectral domain analysis method [29] are also used in EBG characterizations.

Although there exists some challenges [30], numerical methods are still widely used in EBG research because they can analyze general EBG structures with arbitrary configurations. For example, the method of moment (MoM) is implemented in [31] and the finite element method (FEM) is utilized in [32–34]. Time domain technique is also used in EBG analysis [35]. A spectral finite difference time domain (S-FDTD) is developed in [36], where the periodic boundary condition is simply modeled and both visible and surface wave regions can be simulated in a unified approach.

2.3 Interesting EBG properties

EBG structures exhibit exciting electromagnetic properties, such as in-phase reflection and surface wave suppression [37]. The electromagnetic scattering properties of EBG structures, including both the far field radar cross section (RCS) and near field source excitation are investigated in several papers [38–43]. The EBG in-phase reflection is characterized in [44] with a detailed parametric study and the effects of oblique incidences are investigated in [37, 45]. The richness of dispersion relations is another interesting topic in EBG research [46]. The modal characteristics are characterized, and surface mode, leaky mode, and radiation mode are identified in [47–49].

EBG structures share some similar properties with other engineered materials, such as soft/hard surface and double negative materials. For example, [50–51] calculate the soft and hard factors of the EBG surface for TE and TM waves. [52–53] compare band gap features between EBG and soft surfaces. The relation between EBG and left/right handed structures are investigated in [54]. Some fundamental limits on 2-D EBG substrates are discussed in [55].

3 EBG designs

3.1 New materials and configurations

Since the late 1990s, EBG designs have flourished, and a wide variety of materials and geometries have been investigated. In general, EBG structures are periodic composite materials where metal and dielectric materials are arranged in a specific pattern [19, 56–58]. Bi-anisotropic media [59] and ferrite substrate [60–63] are also used in EBG design. Some new fabrication methods have been developed [64], and a representative example is the low temperature co-fired ceramic (LTCC) technique [65].

Many new EBG geometries have been presented, such as the dumbbell shape [66–68], cylindrical, and elliptical configurations [69–73]. Convoluted elements [74], 2-D gyro-tropic EBG [75], and double-arm generic microstrip design [76] are also studied. In addition, multi-layer structure is used to design EBG [16, 77–78]. Since frequency selective surfaces (FSS) are well-known periodic structures, they have led to many novel EBG designs [79–85].

3.2 Enhanced EBG performance

These new EBG designs exhibit improved electromagnetic performance. One important category is wideband EBG designs [86–87]. Multi-period cell is one effective method [88–89] while some other researchers use circuit loading methods such as negative impedance loading [90] and resonant circuit loading [91]. In addition to a single wideband, dual band and multi-band EBG designs are also proposed [92–95].

Compact EBG size is another goal for many EBG designs, which has attracted a broad attention from EBG researchers [91, 95–101]. Several miniaturization techniques are proposed, including the complementary geometries [102–103], spiral shape EBG designs [104–106], and Hilbert curve inclusion [107].

EBG surfaces with polarization-dependent reflection phases are also proposed [108–111]. Using asymmetric geometries such as rectangular shape or offset vias, the reflection phase varies with the polarization of the incident plane wave. They can be used to design polarization converters. Furthermore, reconfigurable and tunable EBG surfaces are studied by different research groups [112–120]. By implementing varactors or other active devices, the EBG properties can be smartly controlled, which is used for antenna beam steering and ultra-thin absorbers.

With the increasing complexity in EBG geometries and challenges in EBG properties, optimization techniques are used to design optimum EBG structures. Two popular techniques have been found in EBG designs, namely, genetic algorithm (GA) [121–123] and particle swarm optimization [124–125].

4 EBG applications in antenna engineering

Because of the unique electromagnetic properties, EBG structures have been widely considered in antennas and microwave circuits to improve their performance. Here we focus on EBG applications in antenna engineering, and typical examples include microstrip antennas, low profile wire antennas, slot antennas, high gain antennas, etc. Several review articles on these applications can be found in conference proceedings and antenna journals [126–134].

4.1 Microstrip antennas and array

The band gap feature of EBG structures has found useful applications in suppressing the surface waves in microstrip antenna designs. As a consequence, the antenna gain and efficiency are increased while the back lobes are reduced. This application has been extensively studied and the results can be found in [135–149]. In particular, several circularly polarized microstrip antennas are investigated in [150–152] and a suspended patch antenna with EBG is explored in [153]. Due to the interactions between EBG and microstrip antennas, several papers reported an improved bandwidth [154–158]. In addition, EBG is used in microstrip antenna designs for size reduction [159] and harmonic control [160]. Another interesting application of EBG is the radiation pattern control for microstrip antennas, as revealed in [161–164].

Besides the antenna element applications, EBG structures have also been used in microstrip antenna arrays. As an important parameter in array design, the mutual coupling between array elements affects the entire array performance such as beam scanning angle. Strong mutual coupling could reduce the array efficiency and cause the scan blindness in phased array systems. Thus, the EBG structures are used to reduce the mutual coupling and eliminate the scan angle, and detailed results can be found in [165–181].

4.2 Low profile wire antennas and slot antennas

Another important group of EBG applications are related to low profile wire antenna designs. Due to the in-phase reflection feature, the radiation efficiency of wire antennas near an EBG ground plane can be improved [182–186]. Different types of wire antennas have been investigated, including dipole antennas [187–190], monopole antennas [16, 191–192], spiral antennas [193–195], curl antennas [196–198], and loop antennas

[199]. Performance of a low profile fractal antenna [200] and an inverted F antenna [201] on the EBG ground plane are also evaluated.

Broadband wire antennas are of special interest to researchers. The Archimedean spiral antenna on the EBG ground plane is revealed to have a broad bandwidth with a low profile configuration [202–203]. Open-sleeve and bow-tie dipole antennas are also probed for wideband operation [204–207].

Low profile wire antenna arrays are studied in several papers. Mutual coupling reduction is reported in [208–210]. Spiral antenna arrays are characterized in [211–212] while a curl array is presented in [213]. Furthermore, an inverted F array is discussed with a tilted beam [214–215].

EBG structures have also been used to improve the radiation performance of slot antennas. For example, they are used in [216–217] to reduce the parallel plate modes. The radiation enhancements on printed slot antennas are reported in [218–220]. A tri-band slot antenna is discussed in [221–222]. Furthermore, EBG applications in waveguide slot arrays are investigated in [223–226].

4.3 High gain antennas

In high gain antenna designs, EBG structures also found several useful applications [227–232]. Using EBG as a superstrate, a new type of high gain antenna referred to as an EBG resonator antenna has been developed [233–238]. Woodpile EBG is popularly used in these designs. EBG resonator antennas with circular polarization [239–240] and improved bandwidth [241–242] are also presented. Furthermore, EBG superstrate for dual polarized spare array is discussed in [243].

Horn antenna is another application area of EBG structures. Quasi-TEM antennas with EBG surfaces are investigated in [244–245] while a woodpile EBG sectoral horn is presented in [246–247]. In addition to dielectric EBG structures, metallic EBG with rectangular, cylindrical and elliptical shapes are used to design directive antenna [248]. An EBG application in the reflector antenna feed is proposed in [249]. In [250–256], a number of examples demonstrate the EBG capabilities in adaptive and beam steering antenna designs.

4.4 EBG antennas in real-life applications

EBG assisted antenna designs, which have improved performance of some of the traditional antennas, have been applied in many real-life applications. Both base station antenna and handset antenna examples are presented in [257–259]. EBG antennas for wireless LAN [260] and microwave links [261] are studied as well. Some EBG antennas are designed for GPS applications [262–263].

Radio frequency identification (RFID) is a rapidly developing technology for automatic target identifications. Several EBG antennas are proposed in RFID tags and RFID readers [264–266]. Besides, some EBG antennas are considered in wearable communication systems [267–268] and implantable biotelemetry systems [269–270]. In direction finding [271], penetration radar [272], and mini-UAV [273], EBG antennas also show promising potential.

In addition to the antenna applications mentioned above, EBG structures also find extensive applications in the broad electromagnetic areas. To name a few, these include waveguides [274–275], microwave circuits [276–279], and reduction of electromagnetic interference [280–283]. Readers are encouraged to use this literature review as a guide to find relevant EBG papers related to their own research interests.

References

1. E. Yablonovitch, "Inhibited spontaneous emission in solid-state physics and electronics," *Phys. Rev. Lett.*, **vol. 58**, 2059–63, 1987.
2. S. John, "Strong localization of photons in certain disordered dielectric super lattices," *Phys. Rev. Lett.*, **vol. 58**, 2486–9, 1987.
3. J. D. Joannopoulos, R. D. Meade, and J. N. Winn, *Photonic Crystals*, Princeton University Press, 1995.
4. C. M. Bowden, J. P. Dowling, and H. O. Everitt, "Development and applications of materials exhibiting photonic band gaps," *J. Opt. Soc. Amer B., Opt. Phys.*, **vol. 10**, 280, February 1993.
5. G. Kurizki and J. W. Haus, "Special issue on photonic band structures," *J. Mod. Opt.*, **vol. 41**, 171–2, February 1994.
6. *IEEE Trans. Microwave Theory Tech., Special Issue on Electromagnetic Crystal Structures, Designs, Synthesis, and Applications*, **vol. 47, no. 11**, November 1999.
7. K. M. Ho, C. T. Chan, and C. M. Soukoulis, "Existence of a photonic gap in periodic dielectric structures," *Phys. Rev. Lett.*, **vol. 65**, 3152–5, 1990.
8. E. Yablonovitch, T. J. Gmitter, and K. M. Leung, "Photonic band structures: the face-centered-cubic case employing non-spherical atoms," *Phys. Rev. Lett.*, **vol. 67**, 2295–8, 1991.
9. P. K. Kelly, J. G. Maloney, B. L. Shirley, and R. L. Moore, "Photonic bandgap structures of finite thickness: theory and experiment," *Proc. 1994 IEEE APS Int. Symp.*, vol. 2, pp. 718–21, Seattle, WA, June 1994.
10. M. M. Beaky, J. B. Burk, H. O. Everitt, M. A. Haider, and S. Venakides, "Two-dimensional photonic crystal Fabry-Perot resonators with lossy dielectrics," *IEEE Trans. Microwave Theory Tech.*, **vol. 47, no. 11**, 2085–91, 1999.
11. K. M. Ho, C. T. Chan, C. M. Soukoulis, R. Biswas, and M. Sigalis, "Photonic band gaps in three dimensions: new layer-by-layer periodic structures," *Solid State Commun.*, **vol. 89**, 413–16, 1994.
12. T. K. Wu (Ed.), *Frequency Selective Surface and Grid Array*, John Wiley & Sons, Inc., 1995.
13. B. A. Munk, *Frequency Selective Surfaces: Theory and Design*, John Wiley & Sons, Inc., 2000.
14. P.-S. Kildal, "Artificial soft and hard surfaces in electromagnetics," *IEEE Trans. Antennas Propagat.*, **vol. 38, no. 10**, 1537–44, 1990.
15. D. Sievenpiper, L. Zhang, R. F. J. Broas, N. G. Alexopolus, and E. Yablonovitch, "High-impedance electromagnetic surfaces with a forbidden frequency band," *IEEE Trans. Microwave Theory Tech.*, **vol. 47**, 2059–74, 1999.

16. D. F. Sievenpiper, *High-Impedance Electromagnetic Surfaces*, Ph.D. Dissertation, Electrical Engineering Dept., University of California, Los Angeles, 1999.
17. F.-R. Yang, K.-P. Ma, Y. Qian, and T. Itoh, "A uniplanar compact photonic-bandgap (UC-PBG) structure and its applications for microwave circuits," *IEEE Trans. Microwave Theory Tech.*, **vol. 47, no. 8**, 1509–14, 1999.
18. F.-R. Yang, *Novel Periodic Structures for Applications to Microwave Circuits*, Ph.D. Dissertation, Electrical Engineering Dept., University of California, Los Angeles, 1999.
19. Y. Rahmat-Samii and H. Mosallaei, "Electromagnetic band-gap structures: classification, characterization and applications," *Proceeding of IEE-ICAP symposium*, pp. 560–4, April 2001.
20. A. Dellavilla, V. Galdi, F. Capolino, V. Pierro, S. Enoch, and G. Tayeb, "A Comparative Study of Representative Categories of EBG Dielectric Quasi-Crystals," *IEEE Antennas Wireless Propagat. Lett.*, **vol. 5**, 331–4, 2006.
21. S. Shahparnia and O. M. Ramahi, "Simple and accurate circuit models for high-impedance surfaces embedded in printed circuit boards," *IEEE APS Int. Symp. Dig.*, vol. 4, pp. 3565–8, June 2004.
22. H. Kim and R. F. Drayton, "Development of analysis method of electromagnetic bandgap (EBG) structures to predict EBG behavior based on circuit models," *IEEE APS Int. Symp. Dig.*, pp. 1951–4, July 2006.
23. M. Rahman and M. A. Stuchly, "Transmission line – periodic circuit representation of planar microwave photonic bandgap structures," *Microwave Optical Tech. Lett.*, **vol. 30, no. 1**, 15–19, 2001.
24. X. H. Wu, A. A. Kishk, and A. W. Glisson, "A transmission line method to compute far-field radiation by arbitrarily directed Hertzian dipoles in a 1-D EBG structure," *IEEE APS Int. Symp. Dig.*, pp. 2997–3000, July 2006.
25. L. Li, B. Li, H.-X. Liu, and C.-H. Liang, "Locally resonant cavity cell model for electromagnetic band gap structures," *IEEE Trans. Antennas Propagat.*, **vol. 54, no. 1**, 90–100, 2006.
26. A. Cucini, M. Caiazzo, M. Nannetti, and S. Maci, "A network theory for EBG surfaces: generalization to any direction of propagation in the azimuth plane," *IEEE APS Int. Symp. Dig.*, vol. 3, pp. 2564–7, June 2004.
27. C. Gao and Y. Wang, "Analysis of EBG structures implemented on CPW components by using EM-ANN models," *IEEE APS Int. Symp. Dig.*, vol. 4, pp. 368–71, June 2002.
28. K. Brakora, C. Barth, and K. Sarabandi, "A plane-wave expansion method for analyzing propagation in 3D periodic ceramic structures," *IEEE APS Int. Symp. Dig.*, vol. 2B, pp. 192–5, July 2005.
29. Z. Sipus, P.-S. Kildal, and R. Zentner, "Spectral domain analysis of dipole coupling over different electromagnetic band gap structures," *IEEE APS Int. Symp. Dig.*, vol. 1A, pp. 738–41, July 2005.
30. R. Mittra, "Numerical challenges in the modeling of metamaterials and antenna-EBG composites," *IEEE APS Int. Symp. Dig.*, vol. 3B, pp. 10–13, July 2005.
31. M. Bozzi, S. Germani, L. Minelli, L. Perregrini, and P. de Maagt, "Full-wave characterization of planar EBG structures by the MoM/BI-RME method," *IEEE APS Int. Symp. Dig.*, vol. 4, pp. 4056–9, June 2004.
32. L. Zhang, *Numerical Characterization of Electromagnetic Band-Gap Materials and Applications in Printed Antennas and Arrays*, Ph.D. Dissertation, Electrical Engineering Dept., University of California, Los Angeles, 2000.

33. X. Zhao and L. Zhou, "Study on 2-D gyrotropic EBG by FEM," *IEEE Int. Symp. Microw. Antennas Propagat. EMC*, vol. 1, pp. 827–30, August 2005.
34. M. N. Vouvakis, Z. Cendes, and J.-F. Lee, "A FEM domain decomposition method for photonic and electromagnetic band gap structures," *IEEE Trans. Antennas Propagat.*, **vol. 54, no. 2, part 2**, 721–33, February 2006.
35. N. Bushyager, J. Papapolymerou, and M. M. Tentzeris, "A composite cell multi-resolution time-domain technique for the design of antenna systems including electromagnetic band gap and via-array structures," *IEEE Trans. Antennas Propagat.*, **vol. 53, no. 8**, 2700–10, August 2005.
36. A. Aminian, F. Yang, and Y. Rahmat-Samii, "Bandwidth determination for soft and hard ground planes by spectral FDTD: a unified approach in visible and surface wave regions," *IEEE Trans. Antennas Propagat.*, **vol. 53, no. 1**, 18–28, January 2005.
37. A. Aminian, F. Yang, and Y. Rahmat-Samii, "In-phase reflection and EM wave suppression characteristics of electromagnetic band gap ground planes," *IEEE APS Int. Symp. Dig.*, vol. 4, pp. 430–3, 22–27 June 2003.
38. W. M. Merrill, C. A. Kyriazidou, H. F. Contopanagos, and N. G. Alexopoulos, "Electromagnetic scattering from a PBG material excited by an electric line source," *IEEE Trans. Microwave Theory Tech.*, **vol. 47, no. 11**, 2105–14, November 1999.
39. C.-S. Kee, J.-E. Kim, H. Y. Park, and H. Lim, "Roles of wave impedance and refractive index in photonic crystals with magnetic and dielectric properties," *IEEE Trans. Microwave Theory Tech.*, **vol. 47, no. 11**, 2148–50, November 1999.
40. H. Jia and K. Yasumoto, "A new analysis of electromagnetic scattering from two-dimensional electromagnetic band-gap structures," *Proceedings, ICCEA 2004*, pp. 21–4, 1–4 November 2004.
41. F. Capolino, D. R. Jackson, and D. R. Wilton, "Fundamental properties of source-excited field at the interface of a 2D EBG material," *IEEE APS Int. Symp. Dig.*, vol. 2, pp. 1171–4, June 2004.
42. F. Capolino, D. R. Jackson, and D. R. Wilton, "Fundamental properties of the field at the interface between air and a periodic artificial material excited by a line source," *IEEE Trans. Antennas Propagat.*, **vol. 53**, 91–9, January 2005.
43. T. Bertuch, "Comparative Investigation of Coupling Reduction by EBG Surfaces for Quasi-Static RCS Measurement Systems," *IEEE Antennas Wireless Propagat. Lett.*, **vol. 5**, 231–4, December 2006.
44. F. Yang and Y. Rahmat-Samii, "Reflection phase characterizations of the EBG ground plane for low profile wire antenna applications," *IEEE Trans. Antennas Propagat.*, **vol. 51, no. 10**, 2691–703, October 2003.
45. L. Li, X. Dang, L. Wang, B. Li, H. Liu, and C. Liang, "Reflection phase characteristics of plane wave oblique incidence on the mushroom-like electromagnetic band-gap structures," *Proceedings of Asia Pacific Microwave Conf.*, December 2005.
46. S. Enoch, G. Tayeb, and B. Gralak, "The richness of the dispersion relation of electromagnetic bandgap materials," *IEEE Trans. Antennas Propagat.*, **vol. 51, no. 10**, 2659–66, 2003.
47. Y.-C. Chen, C.-K. C. Tzuang, T. Itoh, and T. K. Sarkar, "Modal characteristics of planar transmission lines with periodical perturbations: their behaviors in bound, stopband, and radiation regions," *IEEE Trans. Antennas Propagat.*, **vol. 53, no. 1, part 1**, 47–58, 2005.
48. P. Baccarelli, C. Di Nallo, S. Paulotto, and D. R. Jackson, "A full-wave numerical approach for modal analysis of 1-D periodic microstrip structures," *IEEE Trans. Microw. Theory and Tech.*, **vol. 54, no. 4, part 1**, 1350–62, 2006.

49. P. Baccarelli, S. Paulotto, and C. D. Nallo, "Full-wave analysis of bound and leaky modes propagating along 2D periodic printed structures with arbitrary metallisation in the unit cell," *IET Proc. Microwave Antennas Propagation*, **vol. 1, no. 1**, 217–25, 2007.
50. A. Aminian and Y. Rahmat-Samii, "Bandwidth determination for soft and hard ground planes: a unified approach in visible and surface wave regions," *IEEE APS Int. Symp. Dig.*, vol. 1, pp. 313–16, June 2004.
51. A. Aminian, Fan Yang, and Y. Rahmat-Samii, "Bandwidth determination for soft and hard ground planes by spectral FDTD: a unified approach in visible and surface wave regions," *IEEE Trans. Antennas Propagat.*, **vol. 53, no. 1**, 18–28, 2005.
52. E. Rajo-Iglesias, P.-S, Kildal, M. Caiazzo, and J. Yang, "Comparison between bandgaps and bandwidths of back radiation of different narrow soft ground planes," *IEEE APS Int. Symp. Dig.*, vol. 1A, pp. 697–700, July 2005.
53. E. Rajo-Iglesias, M. Caiazzo, L. Inclan-Sanchez, and P.-S. Kildal, "Comparison of bandgaps of mushroom-type EBG surface and corrugated and strip-type soft surfaces," *IET Proc. Microwave Antennas Propagation*, **vol. 1, no. 1**, 184–9, 2007.
54. C. Gao, Z. N. Chen, Y. Y. Wang, and N. Yang, "Investigation on relationship of electromagnetic bandgap structures and left/right handed structures," *2005 IEEE International Workshop on Antenna Technology: Small Antennas and Novel Metamaterials*, pp. 387–90, March 2005.
55. W. H. She, X. Gong, and W. J. Chappell, "Fundamental constraints on two-dimensional EBG substrates," *IEEE APS Int. Symp. Dig.*, vol. 2, pp. 1419–22, June 2004.
56. W. Chappell and L. P. B. Katehi, "Composite metamaterial systems for two-dimensional periodic structures," *IEEE APS Int. Symp. Dig.*, vol. 2, pp. 384–7, June 2002.
57. H. Mosallaei and K. Sarabandi, "Periodic meta-material structures in electromagnetics: concept, analysis, and applications," *IEEE APS Int. Symp. Dig.*, vol. 2, pp. 380–3, June 2002.
58. A. P. Feresidis, G. Goussetis, and J. C. Vardaxoglou, "Metallodielectric arrays without vias as artificial magnetic conductors and electromagnetic band gap surfaces," *IEEE Trans. Antennas Propagat.*, **vol. 2**, 1159–62, 2004.
59. L. G. Zheng and W. X. Zhang, "Research on EBG structure consisting of bianisotropic media," *IEEE APS Int. Symp. Dig.*, vol. 4, pp. 4519–22, June 2004.
60. D. J. Kern and D. H. Werner, "The synthesis of metamaterial ferrites for RF applications using electromagnetic bandgap structures," *IEEE APS Int. Symp. Dig.*, vol. 1, pp. 497–500, June 2003.
61. D. Kim, M. Kim, and S.-W. Kim, "A microstrip phase shifter using ferroelectric electromagnetic bandgap ground plane," *IEEE APS Int. Symp. Dig.*, vol. 2, pp. 1175–8, June 2004.
62. D. J. Kern, D. H. Werner, and M. Lisovich, "Metaferrites: using electromagnetic bandgap structures to synthesize metamaterial ferrites," *IEEE Trans. Antennas Propagat.*, **vol. 53, no. 4**, 1382–9, 2005.
63. J. M. Bell, M. F. Iskander, and J. J. Lee, "Ultrawideband hybrid EBG/ferrite ground plane for low-profile array antennas," *IEEE Trans. Antennas Propagat.*, **vol. 55, no. 1**, 4–12, 2007.
64. Y. Pang, B. Gao, and Z. Du, "Novel electromagnetic bandgap structure fabricating method," *IEEE APS Int. Symp. Dig.*, vol. 4, pp. 372–5, June 2002.
65. X. Gong, T. Smyth, and W. J. Chappell, "Cofiring different dielectric constants inside LTCC for metamaterial applications," *IEEE APS Int. Symp. Dig.*, pp. 1935–8, 9–14 July 2006.

66. A. Yu and X. Zhang, "A novel 2-D electromagnetic band-gap structure and its application in micro-strip antenna arrays," *Proceedings of ICMMT 2002*, pp. 580–3, 17–19 August 2002.
67. A. Yu and X. Zhang, "A novel method to improve the performance of microstrip antenna arrays using a dumbbell EBG structure," *Antennas Wireless Propagat. Lett.*, **vol. 2, no. 1**, 170–2, 2003.
68. A. Yu and X. Zhang, "A low profile monopole antenna using a dumbbell EBG structure," *IEEE APS Int. Symp. Dig.*, vol. 2, pp. 1155–8, 20–25 June 2004.
69. G. K. Palikaras, A. P. Feresidis, and J. C. Vardaxoglou, "Cylindrical electromagnetic bandgap structures for directive base station antennas," *IEEE Antennas Wireless Propagat. Lett.*, **vol. 3, no. 1**, 87–9, 2004.
70. G. K. Palikaras, A. P. Feresidis, and J. C. Vardaxoglou, "Cylindrical electromagnetic band gap structures for base station antennas," *IEEE APS Int. Symp. Dig.*, vol. 2, pp. 1163–6, June 2004.
71. N. Llombart, A. Neto, G. Gerini, and P. de Maagt, "Planar circularly symmetric EBG structures for reducing surface waves in printed antennas," *IEEE Trans. Antennas Propagat.*, **vol. 53, no. 10**, 3210–18, October 2005.
72. H. Boutayeb, T. A. Denidni, A. R. Sebak, and L. Talbi, "Design of elliptical electromagnetic bandgap structures for directive antennas," *IEEE Antennas Wireless Propagat. Lett.*, **vol. 4**, 93–6, 2005.
73. H. Boutayeb, T. A. Denidni, K. Mahdjoubi, A.-C. Tarot, A.-R. Sebak, and L. Talbi, "Analysis and design of a cylindrical EBG-based directive antenna," *IEEE Trans. Antennas Propagat.*, **vol. 54, no. 1**, 211–19, 2006.
74. S. Tse, B. S. Izquierdo, J. C. Batchelor, and R. J. Langley, "Convoluted elements for electromagnetic band gap structures," *IEEE APS Int. Symp. Dig.*, vol. 1, pp. 819–22, June 2004.
75. X. Zhao and L. Zhou, "Study on 2-D gyrotropic EBG by FEM," *IEEE Int. Symp. Microw. Antennas Propagat. EMC*, vol. 1, pp. 827–30, August 2005.
76. G. Cakir and L. Sevgi, "A double-arm generic microstrip electromagnetic bandgap structure with bandpass and bandstop characteristics," *5th Int. Conf. Antenna Theory Techniques*, pp. 464–6, May 2005.
77. N. Boisbouvier, A. Louzir, F. Le Bolzer, A.-C. Tarot, and K. Mahdjoubi, "A double layer EBG structure for slot-line printed devices," *IEEE APS Int. Symp. Dig.*, vol. 4, pp. 3553–6, June 2004.
78. A. P. Feresidis, G. Apostolopoulos, N. Serfas, and J. C. Vardaxoglou, "Closely coupled metallodielectric electromagnetic band-gap structures formed by double-layer dipole and tripole arrays," *IEEE Trans. Antennas Propagat.*, **vol. 52, no. 5**, 1149–58, 2004.
79. J. Yeo, R. Mittra, and S. Chakravarty, "A GA-based design of electromagnetic bandgap (EBG) structures utilizing frequency selective surfaces for bandwidth enhancement of microstrip antennas," *IEEE APS Int. Symp. Dig.*, vol. 2, pp. 400–3, 16–21 June 2002.
80. S. Maci, M. Caiazzo, A. Cucini, and M. Asaletti, "A pole-zero matching method for EBG surfaces composed of a dipole FSS printed on a grounded dielectric slab," *IEEE Trans. Antennas Propag.*, **vol. 53, no. 1, part 1**, 70–81, 2005.
81. G. Goussetis, A. P. Feresidis, and J. C. Vardaxoglou, "FSS printed on grounded dielectric substrates: resonance phenomena, AMC and EBG characteristics," *IEEE APS Int. Symp. Dig.*, vol. 1B, pp. 644–7, July 2005.
82. Y. J. Lee, J. Yeo, R. Mittra, and W. S. Park, "Design of a frequency selective surface (FSS) type superstrate for dual-band directivity enhancement of microstrip patch antennas," *IEEE APS Int. Symp. Dig.*, vol. 3A, pp. 2–5, July 2005.

83. G. Goussetis, A. P. Feresidis, and J. C. Vardaxoglou, "Tailoring the AMC and EBG characteristics of periodic metallic arrays printed on grounded dielectric substrate," *IEEE Trans. Antennas Propagat.*, **vol. 54, no. 1**, 82–9, 2006.
84. E. Rodes, M. Diblanc, J. Drouet, M. Thevenot, T. Monediere, and B. Jecko, "Design of a dual-band EBG resonator antenna using capacitive FSS," *IEEE APS Int. Symp. Dig.*, pp. 3009–12, 9–14 July 2006.
85. A. Pirhadi, F. Keshmiri, and M. Hakkak, "Design of dual-band low profile high directive EBG resonator antenna, using single layer frequency selective surface (FSS) superstrate," *IEEE APS Int. Symp. Dig.*, pp. 3005–8, July 2006.
86. W. J. Chappell and X. Gong, "Wide bandgap composite EBG substrates," *IEEE Trans. Antennas Propagat.*, **vol. 51, no. 10**, 2744–50, 2003.
87. J. M. Bell, M. F. Iskander, and J. J. Lee, "Ultrawideband hybrid EBG/ferrite ground plane for low-profile array antennas," *IEEE Trans. Antennas Propagat.*, **vol. 55, no. 1**, 4–12, 2007.
88. C. C. Chiau, X. Chen, and C. Parini, "Multiperiod EBG structure for wide stopband circuits," *IEE Proc. Microw. Antennas Propagat.*, **vol. 150, no. 6**, 489–92, 2003.
89. Y-C. Chen, A.-S. Liu, and R.-B. Wu, "A wide-stopband low-pass filter design based on multi-period taper-etched EBG structure," *Proceedings of 2005 Asia Pacific Microwave Conf.*, vol. 3, December 2005.
90. D. J. Kern, D. H. Werner, and M. J. Wilhelm, "Active negative impedance loaded EBG structures for the realization of ultra-wideband Artificial Magnetic Conductors," *IEEE APS Int. Symp. Dig.*, vol. 2, pp. 427–30, 22–27 June 2003.
91. H. Mosallaei and K. Sarabandi, "A compact wide-band EBG structure utilizing embedded resonant circuits," *IEEE Antennas Wireless Propagat. Lett.*, **vol. 4**, 5–8, 2005.
92. M. G. Bray and D. H. Werner, "A novel design approach for an independently tunable dual-band EBG AMC surface," *IEEE APS Int. Symp. Dig.*, vol. 1, pp. 289–92, June 2004.
93. D. J. Kern, D. H. Werner, A. Monorchio, L. Lanuzza, and M. J. Wilhelm, "The design synthesis of multiband artificial magnetic conductors using high impedance frequency selective surfaces," *IEEE Trans. Antennas Propagat.*, **vol. 53, no. 1**, 8–17, 2005.
94. Y. Yao, X. Wang, and Z. Feng, "A novel dual-band compact electromagnetic bandgap (EBG) structure and its application in multi-antennas," *IEEE APS Int. Symp. Dig.*, pp. 1943–6, 2006.
95. G. Goussetis, Y. Guo, A. P. Feresidis, and J. C. Vardaxoglou, "Miniaturised and multiband artificial magnetic conductors and electromagnetic band gap surfaces," *IEEE APS Int. Symp. Dig.*, vol. 1, pp. 293–6, June 2005.
96. S. Tse, B. S. Izquierdo, J. C. Batchelor, and R. J. Langley, "Reduced sized cells for high impedance (HIP) ground planes," *Proceedings of 2003 ICAP*, vol. 2, pp. 473–6, 2003.
97. L. Yang and Z. Feng, "Advanced methods to improve compactness in EBG design and utilization," *IEEE APS Int. Symp. Dig.*, vol. 4, pp. 3585–8, June 2004.
98. L. Yang, Z. Feng, F. Chen, and M. Fan, "A novel compact electromagnetic band-gap (EBG) structure and its application in microstrip antenna arrays," *Microwave Symp. Dig.*, vol. 3, pp. 1635–8, June 2004.
99. G. S. A. Shaker and S. Safavi-Naeini, "Reduced size electromagnetic band gap (EBG) structures for antenna applications," *2005 Canadian Conference on Electrical and Computer Engineering*, pp. 1198–201, May 2005.
100. G. S. A. Shaker and S. Safavi-Naeini, "A novel approach for designing miniaturized artificial magnetic conductors (AMCs) and electromagnetic band gap structures (EBGs)," *IEEE APS Int. Symp. Dig.*, vol. 3A, pp. 770–3, July 2005.

101. N. Jin and Y. Rahmat-Samii, "Particle swarm optimization of miniaturized quadrature reflection phase structure for low-profile antenna applications," *IEEE APS Int. Symp. Dig.*, vol. 2B, pp. 255–8, July 2005.
102. G. Apostolopoulos, A. Feresidis, and J. C. Vardaxoglou, "Miniaturised EBG structures based on complementary geometries," *IEEE APS Int. Symp. Dig.*, pp. 2253–6, July 2006.
103. J. C. Vardaxoglou, G. Gousetis, and A. P. Feresidis, "Miniaturisation schemes for metallodielectric electromagnetic bandgap structures," *IET Proc. Microwave Antennas Propagation*, vol. 1, no. 1, pp. 234–9, 2007.
104. C. R. Simovski, P. Maagt, and I. Melchakova, "High-impedance surfaces having stable resonance with respect to polarization and incidence angle," *IEEE Trans. Antennas and Propag.*, **vol. 53, no. 3**, 908–14, 2005.
105. L. Yang, M. Fan, and Z. Feng, "A spiral electromagnetic bandgap (EBG) structure and its application in microstrip antenna arrays," *APMC 2005 Proceedings*, vol. 3, December 2005.
106. Y. Kim, F. Yang, and A. Elsherbeni, "Compact artificial magnetic conductor designs using planar square spiral geometry," *Progress In Electromagnetics Research*, PIER **77**, 43–54, 2007.
107. J. McVay and N. Engheta, "High impedance metamaterial surfaces using Hilbert-curve inclusions," *IEEE Microw. Wireless Components Lett.*, **vol. 14, no. 3**, 130–2, 2004.
108. F. Yang and Y. Rahmat-Samii, "Polarization dependent electromagnetic band-gap surfaces: characterization, designs, and applications," *IEEE APS Int. Symp. Dig.*, vol. 3, pp. 339–42, June 2003.
109. F. Yang and Y. Rahmat-Samii, "Polarization dependent electromagnetic band gap (PDEBG) structures: designs and applications," *Microwave Optical Tech. Lett.*, **vol. 41, no. 6**, 439–44, July 2004.
110. D. Yan, Q. Gao, C. Wang, C. Zhu, and N. Yuan, "Study on polarization characteristic of asymmetrical AMC structure," *Asia Pacific Microw. Conf. Proc.*, December 2005.
111. D. Yan, Q. Gao, C. Wang, C. Zhu, and N. Yuan, "A novel polarization convert surface based on artificial magnetic conductor," *Asia Pacific Microw. Conf. Proc.*, December 2005.
112. D. Sievenpiper, J. Schaffner, B. Loo, G. Tangonan, R. Harold, J. Pikulski, and R. Garcia, "Electronic beam steering using a varactor-tuned impedance surface," *2001 IEEE APS Int. Symp. Dig.*, **vol. 1**, pp. 174–7, July 2001.
113. D. F. Sievenpiper, J. H. Schaffner, H. J. Song, R. Y. Loo, and G. Tangonan, "Two-dimensional beam steering using an electrically tunable impedance surface," *IEEE Trans. Antennas Propagat.*, **vol. 51, no. 10**, 2713–22, October 2003.
114. J. C. Vardaxoglou, A. Chauraya, and P. de Maagt, "Reconfigurable electromagnetic band gap based structures with defects for MM wave and antenna applications," *Proceedings of 2003 ICAP*, vol. 2, pp. 763–6, 2003.
115. D. J. Kern, M. J. Wilhelm, D. H. Werner, and P. L. Werner, "A novel design technique for ultra-thin tunable EBG AMC surfaces," *IEEE APS Int. Symp. Dig.*, vol. 2, pp. 1167–70, 20–25 June 2004.
116. T. Liang, L. Li, J. A. Bossard, D. H. Werner, and T. S. Mayer, "Reconfigurable ultra-thin EBG absorbers using conducting polymers," *IEEE APS Int. Symp. Dig.*, vol. 2B, pp. 204–7, July 2005.
117. L. Mercier, E. Rodes, J. Drouet, L. Leger, E. Arnaud, M. Thevenot, T. Monediere, and B. Jecko, "Steerable and tunable 'EBG resonator antennas' using smart metamaterials," *IEEE APS Int. Symp. Dig.*, pp. 406–9, July 2006.

118. H. Boutayeb and T. A. Denidni, "Technique for reducing the power supply in reconfigurable cylindrical electromagnetic bandgap structures," *IEEE Antennas Wireless Propagat. Lett.*, **vol. 5**, pp. 424–5, December 2006.
119. D. J. Kern, J. A. Bossard, and D. H. Werner, "Design of reconfigurable electromagnetic bandgap surfaces as artificial magnetic conducting ground planes and absorbers," *IEEE APS Int. Symp. Dig.*, pp. 197–200, July 2006.
120. M. G. Bray, Z. Bayraktar, and D. H. Werner, "GA optimized ultra-thin tunable EBG AMC surfaces," *IEEE APS Int. Symp. Dig.*, pp. 410–13, July 2006.
121. J. Yeo, R. Mittra, and S. Chakravarty, "A GA-based design of electromagnetic bandgap (EBG) structures utilizing frequency selective surfaces for bandwidth enhancement of microstrip antennas," *IEEE APS Int. Symp. Dig.*, vol. 2, pp. 400–3, 16–21 June 2002.
122. M. G. Bray, Z. Bayraktar, and D. H. Werner, "GA optimized ultra-thin tunable EBG AMC surfaces," *IEEE APS Int. Symp. Dig.*, pp. 410–13, July 2000.
123. Y. Ge and K. P Esselle, "GA/FDTD technique for the design and optimisation of periodic metamaterials," *IET Proc. Microwave Antennas Propagation*, vol. 1, no. 1, pp. 158–64, 2007.
124. N. Jin and Y. Rahmat-Samii, "Parallel PSO/FDTD algorithm for the optimization of patch antennas and EBG structures," *IEEE/ACES Int. Conf. Wireless Communications and Applied Computational Electromagnetics*, pp. 582–5, April 2005.
125. N. Jin and Y. Rahmat-Samii, "Particle swarm optimization of miniaturized quadrature reflection phase structure for low-profile antenna applications," *IEEE APS Int. Symp. Dig.*, vol. 2B, pp. 255–8, July 2005.
126. F. Yang and Y. Rahmat-Samii, "Applications of electromagnetic band-gap (EBG) structures in microwave antenna designs," *Proceedings, ICMMT 2002*, pp. 528–31, 17–19 August 2002.
127. P. de Maagt, R. Gonzalo, Y. C. Vardaxoglou, and J. M. Baracco, "Electromagnetic bandgap antennas and components for microwave and (sub)millimeter wave applications," *IEEE Trans. Antennas Propagat.*, **vol. 51**, **no. 10**, 2667–77, 2003.
128. D. J. Kern and D. H. Werner, "The synthesis of metamaterial ferrites for RF applications using electromagnetic bandgap structures," *IEEE APS Int. Symp. Dig.*, vol. 1, pp. 497–500, June 2003.
129. Y. Rahmat-Samii, "The marvels of electromagnetic band gap (EBG) structures: novel microwave and optical applications," *Proceedings of the 2003 SBMO/IEEE MTT-S International*, vol. 1, pp. 265–75, 20–23 Sept. 2003.
130. F. Falcone, F. Martin, J. Bonache, T. Lopetegi, M. A. Gomez-Laso, J. Garcia, N. Gil, and M. Sorolla, "Electromagnetic bandgap structures in planar circuit technology," *IEEE APS Int. Symp. Dig.*, vol. 4, pp. 3545–8, June 2004.
131. D. J. Kern, T. G. Spence, and D. H. Werner, "The design optimization of antennas in the presence of EBG AMC ground planes," *IEEE APS Int. Symp. Dig.*, vol. 3A, pp. 10–13, July 2005.
132. J. C. Vardaxoglou, A. P. Feresidis, and G. Goussetis, "Recent advances on EBG and AMC surfaces with applications in terminal and high gain antennas," *7th Int. Conf. Telecommunications in Modern Satellite, Cable and Broadcasting Services*, vol. 1, pp. 3–6, September 2005.
133. R. Zaridze, G. Saparishvili, I. Paroshina, and D. Loskutov, "Computer simulation and experimental fabrication of some EBG antenna devices and electronic circuits," *IEEE APS Int. Symp. Dig.*, vol. 4B, pp. 351–4, July 2005.

134. Y. Lee, Y. Hao, and C. Parini, "Applications of electromagnetic bandgap (EBG) structures for novel communication antenna dsigns," *European Microw. Conf.*, pp. 1056–9, September 2006.
135. R. Gonzalo, P. de Maagt, and M. Sorolla, "Enhanced patch antenna performance by suppressing surface waves using photonic-bandgap substrates," *IEEE Trans. Microwave Theory Tech.*, **vol. 47**, 2131–8, 1999.
136. R. Cocciolo, F. R. Yang, K. P. Ma, and T. Itoh, "Aperture coupled patch antenna on UC-PBG substrate," *IEEE Trans. Microwave Theory Tech.*, **vol. 47**, 2123–30, November 1999.
137. F. Yang, C.-S. Kee, and Y. Rahmat-Samii, "Step-like structure and EBG structure to improve the performance of patch antennas on high dielectric substrate," *IEEE APS Int. Symp. Dig.*, vol. 2, pp. 482–5, 8–13 July 2001.
138. M. Fallah-Rad and L. Shafai, "Enhanced performance of a microstrip patch antenna using a high impedance EBG structure," *IEEE APS Int. Symp. Dig.*, vol. 3, pp. 982–5, June 2003.
139. Y. Zhang, J. von Hagen, M. Younis, C. Fischer, and W. Wiesbeck, "Planar artificial magnetic conductors and patch antennas," *IEEE Trans. Antennas Propagat.*, **vol. 51, no. 10**, 2704–12, 2003.
140. C. C. Chiau, X. Chen, and C. G. Parini, "A microstrip patch antenna on the embedded multi-period EBG structure," *Proceeding of the 6th Int. Symp. Antennas, Propagation and EM Theory*, pp. 96–9, 2003.
141. C. C. Chiau, X. Chen, and C. G. Parini, "A multi-period EBG structure for microstrip antennas," *Proceedings of 2003 ICAP*, vol. 2, pp. 727–30, 2003.
142. N. Llombart, A. Neto, G. Gerini, and P. de Maagt, "Bandwidth, efficiency and directivity enhancement of printed antenna performance using planar circularly symmetric EBGs," *2005 European Microwave Conference*, vol. 3, p. 4, October 2005.
143. N. Llombart, A. Neto, G. Gerini, and P. de Maagt, "Enhanced antenna performances using planar circularly symmetric EBGs," *IEEE APS Int. Symp. Dig.*, vol. 1A, pp. 770–3, July 2005.
144. X. Wang, Y. Hao, and P. S. Hall, "Dual-band resonances of a patch antenna on UC-EBG substrate," *2005 Asia-Pacific Conference Proceedings*, December 2005.
145. S. K. Menon, B. Lethakumary, C. K. Aanandan, K. Vasudevan, and P. Mohanan, "A novel EBG structured ground plane for microstrip antennas," *IEEE APS Int. Symp. Dig.*, vol. 2A, pp. 578–81, July 2005.
146. N. Llombart, A. Neto, G. Gerini, and P. de Maagt, "Planar circularly symmetric EBG structures for reducing surface waves in printed antennas," *IEEE Trans. Antennas Propagat.*, **vol. 53, no. 10**, 3210–18, 2005.
147. R. L. Li, G. DeJean, M. M. Tentzeris, J. Papapolymerou, and J. Laskar, "Radiation-pattern improvement of patch antennas on a large-size substrate using a compact soft-surface structure and its realization on LTCC multilayer technology," *IEEE Trans. Antennas Propagat.*, **vol. 53, no. 1**, 200–8, 2005.
148. W. Gao, "Radiation characteristics of a patch with line-fed on EBG ground," *IEEE APS Int. Symp. Dig.*, pp. 3021–4, July 2006.
149. D. Qu, L. Shafai, and A. Foroozesh, "Improving microstrip patch antenna performance using EBG substrates," *IEE Proc. Microw. Antennas Propag*, vol. 153, no. 6, pp. 558–63, December 2006.
150. T. Sudha and T. S. Vedavathy, "A dual band circularly polarized microstrip antenna on an EBG substrate," *IEEE APS Int. Symp. Dig.*, vol. 2, pp. 68–71, June 2002.
151. M. Rahman and M. A. Stuchly, "Circularly polarised patch antenna with periodic structure," *IEE Proc. Microw. Antennas Propagat.*, vol. 149, no. 3, pp. 141–6, June 2002.

152. J. C. Iriarte, I. Ederra, R. Gonzalo, A. Gosh, J. Laurin, C. Caloz, Y. Brand, M. Gavrilovic, Y. Demers, and P. de Maagt, "EBG superstrate for gain enhancement of a circularly polarized patch antenna," *IEEE APS Int. Symp. Dig.*, pp. 2993–6, 2006.
153. L. Yang, W. Chen, M. Fan, and Z. Feng, "Enhanced performance of a suspended patch antenna with stacked EBG utilization," *IEEE APS Int. Symp. Dig.*, vol. 3, pp. 2412–15, June 2004.
154. J. Yeo, R. Mittra, and S. Chakravarty, "A GA-based design of electromagnetic bandgap (EBG) structures utilizing frequency selective surfaces for bandwidth enhancement of microstrip antennas," *IEEE APS Int. Symp. Dig.*, vol. 2, pp. 400–3, June 2002.
155. G. Kiziltas, D. Psychoudakis, J. L. Volakis, and N. Kikuchi, "Topology design optimization of dielectric substrates for bandwidth improvement of a patch antenna," *IEEE Trans. Antennas Propagat.*, **vol. 51, no. 10**, 2732–43, 2003.
156. D. Qu and L. Shafai, "Wideband microstrip patch antenna with EBG substrates," *IEEE APS Int. Symp. Dig.*, vol. 2A, pp. 594–7, July 2005.
157. D. Qu and L. Shafai, "The performance of microstrip patch antennas over high impedance EBG substrates within and outside its bandgap," *IEEE Int. Symp. Microwave, Antenna, Propagation and EMC Technologies for Wireless Communications*, vol. 1, pp. 423–6, August 2005.
158. L. Leger, T. Monediere, and B. Jecko, "Enhancement of gain and radiation bandwidth for a planar 1-D EBG antenna," *IEEE Microw. Wirel. Co. Lett.*, **vol. 15, no. 9**, 573–5, 2005.
159. S. Pioch and J.-M. Laheurte, "Size reduction of microstrip antennas by means of periodic metallic patterns," *Electron. Lett.*, **vol. 39, no. 13**, 959–61, 2003.
160. Y. Horii and M. Tsutsumi, "Wide band operation of a harmonically controlled EBG microstrip patch antenna," *IEEE APS Int. Symp. Dig.*, **vol. 3**, 768–71, 2002.
161. Y. Lee, J. Yeo, and R. Mittra, "Investigation of electromagnetic bandgap (EBG) structures for antenna pattern control," *IEEE APS Int. Symp. Dig.*, vol. 2, pp. 1115–18, June 2003.
162. Y. J. Lee, J. Yeo, K. D. Ko, R. Mittra, Y. Lee, and W. S. Park, "Techniques for controlling the defect frequencies of electromagnetic bandgap (EBG) superstrates for dual-band directivity enhancement of a patch antenna," *IEEE APS Int. Symp. Dig.*, vol. 2, pp. 1143–6, June 2004.
163. Y. J. Lee, J. Yeo, R. Mittra, and W. S. Park, "Application of electromagnetic bandgap (EBG) superstrates with controllable defects for a class of patch antennas as spatial angular filters," *IEEE Trans. Antennas Propagat.*, **vol. 53, no. 1, part 1**, pp. 224–35, 2005.
164. Y. J. Lee, J. Yeo, R. Mittra, and W. S. Park, "Design of a frequency selective surface (FSS) type superstrate for dual-band directivity enhancement of microstrip patch antennas," *IEEE APS Int. Symp. Dig.*, vol. 3A, pp. 2–5, July 2005.
165. F. Yang and Y. Rahmat-Samii, "Mutual coupling reduction of microstrip antennas using electromagnetic band-gap structure," *IEEE APS Int. Symp. Dig.*, vol. 2, pp. 478–81, July 2001.
166. S.-G. Mao, C.-M. Chen, and D.-C. Chang, "Modeling of slow-wave EBG structure for printed-bowtie antenna array," *IEEE Antennas Wireless Propagat. Lett.*, **vol. 1, no. 1**, 124–7, 2002.
167. A. Yu and X. Zhang, "A novel 2-D electromagnetic band-gap structure and its application in micro-strip antenna arrays," *Proceedings of ICMMT 2002*, pp. 580–3, 17–19 August 2002.
168. F. Yang and Y. Rahmat-Samii, "Microstrip antennas integrated with electromagnetic band-gap (EBG) structures: a low mutual coupling design for array applications," *IEEE Trans. Antennas Propagat.*, **vol. 51, no. 10, part 2**, 2936–46, 2003.

169. A. Yu and X. Zhang, "A novel method to improve the performance of microstrip antenna arrays using a dumbbell EBG structure," *Antennas Wireless Propagat. Lett.*, **vol. 2, no. 1**, 170–2, 2003.
170. M. Bozzetti, A. D'Orazio, M. De Sario, C. Grisorio, V. Petruzzelli, F. Prudenzano, and A. Rotunno, "Electromagnetic bandgap phased array antenna controlled by piezoelectric transducer," *Electron. Lett.*, **vol. 39, no. 14**, 1028–30, July 2003.
171. L. Yang, Z. Feng, F. Chen, and M. Fan, "A novel compact electromagnetic band-gap (EBG) structure and its application in microstrip antenna arrays," *Microwave Symposium Digest*, vol. 3, pp. 1635–8, June 2004.
172. Z. Iluz, R. Shavit, and R. Bauer, "Microstrip antenna phased array with electromagnetic bandgap substrate," *IEEE Trans. Antennas Propagat.*, **vol. 52, no. 6**, 1446–53, 2004.
173. L. C. Kretly and S. A. M. P. Alves, "The effect of an electromagnetic band-gap structure on a PIFA antenna array," *Proceedings of IEEE Symp. Personal, Indoor and Mobile Radio Communications*, vol. 2, pp. 1268–71, September 2004.
174. Y. Fu and N. Yuan, "Elimination of scan blindness in phased array of microstrip patches using electromagnetic bandgap materials," *IEEE Antennas Wireless Propagat. Lett.*, **vol. 3, no. 1**, 63–5, 2004.
175. N. Llombart, A. Neto, G. Gerini, and P. De Maagt, "Planar circularly symmetric EBGs to improve the isolation of array elements," *IEEE APS Int. Symp. Dig.*, vol. 2A, pp. 582–5, 3–8 July 2005.
176. N. Llombart, A. Neto, G. Gerini, and P. de Maagt, "On the use of planar EBGs in one dimensional (1D) scanning printed arrays," *Proceedings of Radar Conf.*, pp. 303–6, October 2005.
177. L. Yang, M. Fan, and Z. Feng, "A spiral electromagnetic bandgap (EBG) structure and its application in microstrip antenna arrays," *2005 Asia-Pacific Conference Proceedings*, December 2005.
178. J. Liang and H. Y. D. Yang, "Analysis of a Proximity Coupled Patch Antenna on a Metallized Substrate," *IEEE APS Int. Symp. Dig.*, pp. 2287–90, July 2006.
179. V. Pynttari, R. Makinen, A. Ruhanen, J. Heikkinen, and M. Kivikoski, "Comparison of electromagnetic band-gap structures for microstrip antenna arrays on thin substrates," *IEEE APS Int. Symp. Dig.*, pp. 3017–20, July 2006.
180. H.-L. Liu, B.-Z. Wang, X.-S. Yang, and W. Shao, "Elimination of mutual couplings in reflectarray using electromagnetic bandgap structures," *IEEE APS Int. Symp. Dig.*, pp. 2299–302, July 2006.
181. G. Donzelli, F. Capolino, S. Boscolo, and M. Midrio, "Elimination of scan blindness in phased array antennas using a grounded-dielectric EBG material," *IEEE Antennas Wireless Propagat. Lett.*, **vol. 6**, 2007.
182. F. Yang and Y. Rahmat-Samii, "Reflection phase characterizations of the EBG ground plane for low profile wire antenna applications," *IEEE Trans. Antennas Propagat.*, **vol. 51, no. 10**, 2691–703, 2003.
183. H. Boutayeb, K. Mahdjoubi, and A. C. Tarot, "Design of a directive and matched antenna with a planar EBG structure," *IEEE APS Int. Symp. Dig.*, vol. 1, pp. 835–8, June 2004.
184. J. R. Sohn, H.-S. Tae, J.-G. Lee, and J.-H. Lee, "Comparative analysis of four types of high-impedance surfaces for low profile antenna applications," *IEEE APS Int. Symp. Dig.*, vol. 1A, pp. 758–61, July 2005.
185. M. F. Abedin and M. Ali, "Effects of EBG reflection phase profiles on the input impedance and bandwidth of ultrathin directional dipoles," *IEEE Trans. Antennas Propagat.*, **vol. 53, no. 11**, 3664–72, 2005.

186. S. Clavijo, R. E. Diaz, and W. E. McKinzie, "Design methodology for Sievenpiper high-impedance surfaces: an artificial magnetic conductor for positive gain electrically small antennas," *IEEE Trans. Antennas Propagat.*, **vol. 51, no. 10**, 2678–90, 2003.
187. M. F. Abedin and M. Ali, "Application of EBG substrates to design ultra-thin wideband directional dipoles," *IEEE APS Int. Symp. Dig.*, vol. 2, pp. 2071–4, June 2004.
188. F. Yang, V. Demir, D. A. Elsherbeni, A. Z. Elsherbeni, and A. A. Eldek, "Planar dipole antennas near the edge of an EBG ground plane for WLAN applications," *IEEE APS Int. Symp. Dig.*, vol. 1A, pp. 750–3, 3–8 July 2005.
189. L. Akhoondzadeh-Asl, P. S. Hall, J. Nourinia, and Ch. Ghobadi, "Influence of Angular Stability of EBG Structures on Low Profile Dipole Antenna Performance," *IEEE Int. Workshop on Antenna Technology Small Antennas and Novel Metamaterials*, pp. 253–6, March, 2006.
190. I. Ederra, R. Gonzalo, B. Martinez, L. Azcona, B. Alderman, P. Huggard, B. P. D. Hon, M. V. Beurden, L. Marchand, and P. de Maagt, "Modifications of the woodpile structure for the improvement of its performance as substrate for dipole antennas," *IET Proc. Microwave Antennas Propagation*, **vol. 1, no. 1**, 226–33, 2007.
191. A. Yu and X. Zhang, "A low profile monopole antenna using a dumbbell EBG structure," *IEEE APS Int. Symp. Dig.*, vol. 2, pp. 1155–8, June 2004.
192. F. Yang and Y. Rahmat-Samii, "Bent monopole antennas on EBG ground plane with reconfigurable radiation patterns," *IEEE APS Int. Symp. Dig.*, vol. 2, pp. 1819–22, June 2004.
193. T. H. Liu, W. X. Zhang, M. Zhang, and K. F. Tsang, "Low profile spiral antenna with PBG substrate," *Electron. Lett.*, **vol. 36, no. 9**, 779–80, 2000.
194. H. Nakano, M. Ikeda, K. Hitosugi, J. Yamauchi, and K. Hirose, "A spiral antenna backed by an electromagnetic band-gap material," *IEEE APS Int. Symp. Dig.*, vol. 4, pp. 482–5, June 2003.
195. H. Nakano, K. Hitosugi, N. Tatsuzawa, D. Togashi, H. Mimaki, and J. Yamauchi, "Effects on the radiation characteristics of using a corrugated reflector with a helical antenna and an electromagnetic band-gap reflector with a spiral antenna," *IEEE Trans. Antennas Propagat.*, **vol. 53, no. 1**, 191–9, 2005.
196. F. Yang and Y. Rahmat-Samii, "Curl antennas over electromagnetic band-gap surface: a low profiled design for CP applications," *IEEE APS Int. Symp. Dig.*, vol. 3, pp. 372–5, July 2001.
197. F. Yang and Y. Rahmat-Samii, "A low profile circularly polarized curl antenna over electromagnetic band-gap (EBG) surface," *Microwave Optical Tech. Lett.*, **vol. 31, no. 4**, 264–7, 2001.
198. P. Raumonen, M. Keskilammi, L. Sydanheimo, and M. Kivikoski, "A very low profile CP EBG antenna for RFID reader," *IEEE APS Int. Symp. Dig.*, vol. 4, pp. 3808–11, June 2004.
199. J. Kim and Y. Rahmat-Samii, "Low-profile loop antenna above EBG structure," *IEEE APS Int. Symp. Dig.*, vol. 2A, pp. 570–3, July 2005.
200. B.-L. Ooi, "A modified contour Integral analysis for Sierpinski fractal carpet antennas with and without electromagnetic band gap ground plane," *IEEE Trans. Antennas Propagat.*, **vol. 52, no. 5**, 1286–93, 2004.
201. H. Nakano, Y. Asano, and J. Yamauchi, "A wire inverted F antenna on a finite-sized EBG material," *IEEE International Workshop on Antenna Technology: Small Antennas and Novel Metamaterials*, pp. 13–16, March 2005.
202. J. M. Bell and M. F. Iskander, "A low-profile Archimedean spiral antenna using an EBG ground plane," *IEEE Antennas Wireless Propagat. Lett.*, **vol. 3, no. 1**, 223–6, 2004.

203. L. Schreider, X. Begaud, M. Soiron, B. Perpere, and C. Renard, "Broadband Archimedean spiral antenna above a loaded electromagnetic band gap substrate," *IET Proc. Microwave Antennas Propagation*, vol. 1, no. 1, pp. 212–16, 2007.
204. M. G. Bray and D. H. Werner, "A broadband open-sleeve dipole antenna mounted above a tunable EBG AMC ground plane," *IEEE APS Int. Symp. Dig.*, vol. 2, pp. 1147–50, June 2004.
205. S. R. Best and D. L. Hanna, "Design of a broadband dipole in close proximity to an EBG ground plane," *2005 Antenna Applications Symposium*, University of Illinois, September 2005.
206. L. Akhoondzadeh-Asl and P. S. Hall, "Wideband dipoles on electromagnetic bandgap ground planes," *IEE Conf. Wideband and Multi-band Antennas and Arrays*, pp. 41–5, September 2005.
207. L. Akhoondzadeh, D. J. Kern, P. S. Hall, and D. H. Werner, "Wideband dipoles on electromagnetic bandgap ground planes," *IEEE Trans. Antennas Propagat.*, **vol. 55, no. 9**, 2426–34, September 2007.
208. M. F. Abedin and M. Ali, "Reducing the mutual-coupling between the elements of a printed dipole array using planar EBG structures," *IEEE APS Int. Symp. Dig.*, vol. 2A, 598–601, 2005.
209. M. F. Abedin and M. Ali, "Effects of a smaller unit cell planar EBG structure on the mutual coupling of a printed dipole array," *IEEE Antennas Wireless Propagat. Lett.*, **vol. 4**, 274–6, 2005.
210. I. Ederra, B. M. Pascual, A. B. Labajos, J. Teniente, R. Gonzalo, and P. de Maagt, "Experimental verification of the reduction of coupling between dipole antennas by using a woodpile substrate," *IEEE Trans. Antennas Propagat.*, **vol. 54, no. 7**, 2105–12, 2006.
211. H. Nakano, K. Hitosugi, and J. Yamauchi, "A spiral antenna array with an electromagnetic band-gap reflector," *IEEE APS Int. Symp. Dig.*, vol. 1, pp. 831–4, June 2004.
212. H. Nakano, K. Hitosugi, P. Huang, H. Mimaki, and J. Yamauchi, "A low-profile spiral antenna array above an EBG reflector," *IEEE APS Int. Symp. Dig.*, vol. 3A, pp. 18–21, July 2005.
213. J.-M. Baracco, M. Paquay, and P. de Maagt, "An electromagnetic bandgap curl antenna for phased array applications," *IEEE Trans. Antennas Propagat.*, **vol. 53, no. 1, part 1**, 173–80, 2005.
214. H. Nakano, Y. Asano, H. Mimaki, and J. Yamauchi, "Tilted beam formation by an array composed of strip inverted F antennas with a finite-sized EBG reflector," *IEEE Int. Symp. Microwave, Antenna, Propagation EMC*, vol. 1, pp. 438–41, August 2005.
215. H. Nakano, Y. Asano, G. Tsutsumi, and J. Yamauchi, "A Low-Profile Inverted F Element Array Backed by an EBG Reflector," *IEEE APS Int. Symp. Dig.*, pp. 2985–8, 2006.
216. J. D. Shumpert, W. J. Chappell, and L. P. B. Katehi, "Parallel-plate mode reduction in conductor-backed slots using electromagnetic bandgap substrates," *IEEE Trans. Microwave Theory Tech.*, **vol. 47, no. 11**, 2099–104, 1999.
217. J. M. Fernandez and M. S. Castaner, "Effect of AMC sidewall structures in parallel plate slot antennas," *IEEE APS Int. Symp. Dig.*, vol. 2B, pp. 659–62, July 2005.
218. A. Neto, N. Llombart, G. Gerini, and P. de Maagt, "On the optimal radiation bandwidth of printed slot antennas surrounded by EBGs," *IEEE Trans. Antennas Propagat.*, **vol. 54, no. 4**, 1074–83, 2006.
219. M. A. Habib, M. Nedil, and T. A. Denidni, "Radiation efficiency enhancement of slot antennas using EBG-pin inclusion," *IEEE APS Int. Symp. Dig.*, pp. 2295–8, July 2006.

220. N. Llombart, A. Neto, G. Gerini, and P. de Maagt, "1-D Scanning Arrays on Dense Dielectrics Using PCS-EBG Technology," *IEEE Trans. Antennas Propagat.*, **vol. 55, no. 1**, 26–35, 2007.
221. D. Xu, B. L. Ooi, and G. Zhao, "A new triple-band slot antenna with EBG feed," *IEEE Int. Symp. Microwave Antennas Propagat. EMC*, vol. 1, pp. 41–4, August 2005.
222. B. L. Ooi, X. D. Xu, and I. Ang, "Triple-band slot antenna with spiral EBG feed," *IEEE Int. Workshop Antenna Technology: Small Antennas and Novel Metamaterials*, pp. 329–32, March 2005.
223. L. Bin, L. Long, and C.-H. Liang, "Waveguide slot array antenna with EBG high-impedance surface structure," *2005 Asia-Pacific Microwave Conference Proceedings*, December 2005.
224. Q.-R. Zheng, G.-H. Zhang, and N.-C. Yuan, "Single ridged waveguide slot phased antenna array integrated with high impedance ground plane," *2005 Asia-Pacific Microwave Conference Proceedings*, December 2005.
225. Q. Gao, G.-H. Zhang, D.-B. Yan, and N.-C. Yuan, "Waveguide slot antenna using high-impedance surface," *2005 Asia-Pacific Microwave Conference Proceedings*, December 2005.
226. L. Li, X.-J. Dang, B. Li, and C.-H. Liang, "Analysis and design of waveguide slot antenna array integrated with electromagnetic band-gap structures," *IEEE Antennas Wireless Propagat. Lett.*, **vol. 5, no. 1**, 111–15, 2006.
227. P. K. Kelly, J. G. Maloney, and G. Smith, "Antenna design with the use of photonic bandgap materials as all-dielectric planar reflector," *Microwave and Optical Tech. Lett.*, **vol. 11**, 169–74, 1996.
228. M. Thevenot, A. Reineix, and B. Jecko, "Directive photonic-bandgap antennas," *IEEE Trans. Microwave Theory Tech.*, **vol. 47, no. 11**, 2115–22, 1999.
229. S. Wang, A. P. Feresidis, G. Goussetis, and J. C. Vardaxoglou, "Low profile highly directive antennas using EBG superstrates and metamaterial ground planes," *IEEE APS Int. Symp. Dig.*, vol. 4B, pp. 335–8, July 2005.
230. B. Jecko, T. Monediere, and L. Leger, "High Gain EBG Resonator Antenna," *18th Int. Conf. Applied Electromagnetics Communications*, pp. 1–3, October 2005.
231. J. C. Vardaxoglou and P. de Maagt, "Recent advances on metamaterials with applications in terminal and high gain array and reflector antennas," *IEEE APS Int. Symp. Dig.*, pp. 423–6, July 2006.
232. Y. Vardaxoglou and F. Capolino, "Review of highly-directive flat-plate antenna technology with metasurfaces and metamaterials," *European Microw. Conf.*, pp. 963–6, September 2006.
233. C. Cheype, C. Serier, M. Thevenot, T. Monediere, A. Reineix, and B. Jecko, "An electromagnetic bandgap resonator antenna," *IEEE Trans. Antennas Propagat.*, **vol. 50, no. 9**, 1285–90, 2002.
234. A. R. Weily, K. Esselle, B. C. Sanders, and T. S. Bird, "Woodpile EBG resonator antenna with double slot feed," *IEEE APS Int. Symp. Dig.*, vol. 2, pp. 1139–42, 20–5 June 2004.
235. A. R. Weily, L. Horvath, K. P. Esselle, B. C. Sanders, and T. S. Bird, "A planar resonator antenna based on a woodpile EBG material," *IEEE Trans. Antennas Propagat.*, **vol. 53, no. 1**, 216–23, 2005.
236. M. Thevenot, J. Drouet, B. Jecko, T. Monediere, L. Leger, L. Freytag, R. Chantalat, and M. Diblanc, "New advancements to exploit the potentialities of the EBG resonator antennas," *IEEE APS Int. Symp. Dig.*, vol. 3A, pp. 22–5, July 2005.
237. G. M. Sardi, G. Donzelli, and F. Capolino, "High directivity at broadside with new radiators made of dielectric EBG materials," *IEEE APS Int. Symp. Dig.*, pp. 373–6, July 2006.

238. Y. Lee, X. Lu, Y. Hao, S. Yang, R. Ubic, J. R. G. Evans, and C. G. Parini, "Directive millimetre-wave antenna based on free formed woodpile EBG structure," *Electron. Lett.*, **vol. 43, no. 4**, 195–6, 2007.

239. M. Diblanc, E. Rodes, E. Arnaud, M. Thevenot, T. Monediere, and B. Jecko, "Circularly polarized metallic EBG antenna," *IEEE Microw. Wirel. Co. Lett.*, **vol. 15, no. 10**, 638–40, 2005.

240. A. R. Weily, K. P. Esselle, T. S. Bird, and B. C. Sanders, "High gain circularly polarised 1-D EBG resonator antenna," *Electron. Lett.*, **vol. 42, no. 18**, 3–4, 2006.

241. A. R. Weily, K. P. Esselle, T. S. Bird, and B. C. Sanders, "High gain antenna with improved radiation bandwidth using dual 1-D EBG resonators and array feed," *IEEE APS Int. Symp. Dig.*, pp. 3–10, July 2006.

242. A. R. Weily, K. P. Esselle, T. S. Bird, and B. C. Sanders, "Dual resonator 1-D EBG antenna with slot array feed for improved radiation bandwidth," *IET Proc. Microwave Antennas Propagation*, vol. 1, no. 1, 198–203, 2007.

243. R. Gardelli, M. Albani, and F. Capolino, "EBG superstrates for dual polarized sparse arrays," *IEEE APS Int. Symp. Digest*, vol. 2A, pp. 586–9, July 2005.

244. S. P. Skobelev and P.-S. Kildal, "Analysis of conical quasi-TEM horn with a hard corrugated section," *IEEE Trans. Antennas Propagat.*, **vol. 51, no. 10**, 2723–31, 2003.

245. S. P. Skobelev and P.-S. Kildal, "Mode-matching modeling of a hard conical quasi-TEM horn realized by an EBG structure with strips and vias," *IEEE Trans. Antennas Propagat.*, **vol. 53, no. 1**, 139–43, 2005.

246. A. R. Weily, K. P. Esselle, B. C. Sanders, and T. S. Bird, "A woodpile EBG sectoral horn antenna," *IEEE APS Int. Symp. Dig.*, vol. 4B, pp. 323–6, July 2005.

247. A. R. Weily, K. P. Esselle, T. S. Bird, and B. C. Sanders, "Linear array of woodpile EBG sectoral horn antennas," *IEEE Trans. Antennas Propagat.*, **vol. 54, no. 8**, 2263–74, 2006.

248. H. Boutayeb, T. A. Denidni, A. Sebak, and L. Talbi, "Metallic EBG structures for directive antennas using rectangular, cylindrical and elliptical shapes," *IEEE APS Int. Symp. Dig.*, vol. 1A, pp. 762–5, July 2005.

249. D. Israel, R. Shavit, and Z. Iluz, "Multi-beam reflector antenna with feeds covered by an EBG structure used as a spatial angular filter," *IEEE APS Int. Symp. Dig.*, pp. 353–6, July 2006.

250. D. F. Sievenpiper, J. H. Schaffner, H. J. Song, R. Y. Loo, and G. Tangonan, "Two-dimensional beam steering using an electrically tunable impedance surface," *IEEE Trans. Antennas Propagat.*, **vol. 51, no. 10**, 2713–22, 2003.

251. L. Leger, T. Monediere, M. Thevenot, and B. Jecko, "Multifrequency and beam steered electromagnetic band gap antennas," *IEEE APS Int. Symp. Dig.*, vol. 2, pp. 1151–4, June 2004.

252. H. Talleb, D. Lautru, and V. Fouad-Hanna, "Analysis of adaptive antenna having electromagnetic band gap metallic structures," *IEEE APS Int. Symp. Dig.*, vol. 2A, pp. 566–9, July 2005.

253. P. Ratajczak, P. Brachat, and J. M. Fargeas, "An adaptive beam steering antenna for mobile communications," *IEEE APS Int. Symp. Dig.*, pp. 418–21, July 2006

254. H. Talleb, D. Lautru, and V. Fouad-Hanna, "Analysis of an electronically adaptive antenna using an EBG metallic structure with inserted localized elements," *36th European Microw. Conf.*, pp. 768–71, September 2006.

255. A. Chauraya, C. Panagamuwa, and J. Vardaxoglou, "Beam scanning antenna with photonically tuned EBG phase shifters," *IEEE APS Int. Symp. Dig.*, pp. 2283–6, July 2006.

256. H. Boutayeb and T. A. Denidni, "New configuration of cylindrical EBG structure for beam switching antennas," *IEEE APS Int. Symp. Dig.*, pp. 2271–4, July 2006.
257. G. K. Palikaras, A. P. Feresidis, and J. C. Vardaxoglou, "Cylindrical electromagnetic bandgap structures for directive base station antennas," *IEEE Antennas Wireless Propagat. Lett.*, **vol. 3, no. 1**, 87–9, 2004.
258. L. Freytag, E. Pointereau, and B. Jecko, "Omnidirectional dielectric electromagnetic band gap antenna for base station of wireless network," *IEEE APS Int. Symp. Dig.*, vol. 1, pp. 815–18, June 2004.
259. G. Goussetis, J. C. Vardaxoglou, and A. P. Feresidis, "Handset antenna performance using flexible MEBG structures," *IEEE International Workshop on Antenna Technology: Small Antennas and Novel Metamaterials*, pp. 55–8, March 2005.
260. G. K. Palikaras, A. P. Feresidis, and J. C. Vardaxoglou, "Cylindrical EBG surfaces for omni-directional wireless LAN antennas," *IEEE APS Int. Symp. Dig.*, vol. 4B, pp. 339–42, July 2005.
261. R. Alkhatib and M. Drissi, "EBG antenna for microwave links applications," *Int. Conf. Information and Communication Technologies*, vol. 2, pp. 2190–4, April 2006.
262. X. L. Bao, G. Ruvio, M. J. Ammann, and M. John, "A novel GPS patch antenna on a fractal hi-impedance surface substrate," *IEEE Antennas Wireless Propagat. Lett.*, **vol. 5**, 323–6, 2006.
263. M. Martinez-Vazquez and R. Baggen, "Characterisation of printed EBG surfaces for GPS applications," *IEEE Int. Workshop on Antenna Technology Small Antennas and Novel Metamaterials*, pp. 5–8, March 2006.
264. L. Ukkonen, L. Sydanheimo, and M. Kivikoski, "Patch antenna with EBG ground plane and two-layer substrate for passive RFID of metallic objects," *IEEE APS Int. Symp. Dig.*, vol. 1, pp. 93–6, June 2004.
265. P. Raumonen, M. Keskilammi, L. Sydanheimo, and M. Kivikoski, "A very low profile CP EBG antenna for RFID reader," *IEEE APS Int. Symp. Dig.*, vol. 4, pp. 3808–11, June 2004.
266. M. Stupf, R. Mittra, J. Yeo, and J. R. Mosig, "Some novel design for RFID antennas and their performance enhancement with metamaterials," *IEEE APS Int. Symp. Dig.*, pp. 1023–6, July 2006.
267. P. Salonen, F. Yang, Y. Rahmat-Samii, and M. Kivikoski, "WEBGA – wearable electromagnetic band-gap antenna," *IEEE APS Int. Symp. Dig.*, vol. 1, pp. 451–4, June 2004.
268. S. Zhu and R. Langley, "Dual-band wearable antennas over EBG substrate," *Electron. Lett.*, **vol. 43, no. 3**, 141–2, 2007.
269. J. Kim and Y. Rahmat-Samii, "Exterior antennas for wireless medical links: EBG backed dipole and loop antennas," *IEEE APS Int. Symp. Dig.*, vol. 2B, pp. 800–3, 3–8 July 2005.
270. J. Kim and Y. Rahmat-Samii, "Electromagnetic interactions between biological tissues and implantable biotelemetry systems," *IEEE MTT-S Int. Microwave Symposium Digest*, June 2005.
271. A. Hirata, "Accuracy compensation in direction finding using patch antenna array with EBG structure," *IEEE Antennas Wireless Propagat. Lett.*, **vol. 5**, 1–3, 2006.
272. J. M. Bell, M. F. Iskander, and J. J. Lee, "Ultra-wideband and low-profile hybrid EBG/ferrite ground plane for airborne foliage penetrating radar," *IEEE APS Int. Symp. Dig.*, pp. 369–72, July 2006.
273. P. Salonen and K. Rintala, "An S-band EBG antenna for mini-UAV," *IEEE APS Int. Symp. Dig.*, pp. 2373–6, July 2006.

274. F.-R. Yang, K.-P. Ma, Y. Qian, and T. Itoh, "A novel TEM waveguide using uniplanar compact photonic-bandgap (UC-PBG) structure," *IEEE Trans. Microwave Theory Tech.*, **vol. 47, no. 11**, 2092–8, 1999.
275. T. Kamgaing and O. M. Ramahi, "Electromagnetic band-gap structures for multiband mitigation of resonant modes in parallel-plate waveguides," *IEEE APS Int. Symp. Dig.*, vol. 4, pp. 3577–80, June 2004.
276. S. K. Padhi and N. C. Karmakar, "Spurious harmonics suppression of tapered SIR band-pass filter using electromagnetic bandgap (EBG) structure," *IEEE APS Int. Symp. Dig.*, vol. 4, pp. 3561–4, June 2004.
277. I. Ederra, L. Azcona, B. Alderman, A. Laisne, R. Gonzalo, and P. de Maagt, "A 250 GHz sub-harmonic mixer design implemented in EBG technology," *IEEE APS Int. Symp. Dig.*, vol. 3A, pp. 35–8, July 2005.
278. M. F. Karim, A.-Q. Liu, A. Alphones, X. J. Zhang, and A. B. Yu, "CPW band-stop filter using unloaded and loaded EBG structures," *IEE Proc. Microw. Antennas Propagat.*, pp. 434–40, December 2005.
279. J. J. Simpson, A. Taflove, J. A. Mix, and H. Heck, "Advances in hyperspeed digital interconnects using electromagnetic bandgap technology: measured low-loss 43-GHz passband centered at 50 GHz," *IEEE APS Int. Symp. Dig.*, vol. 3A, pp. 26–9, July 2005.
280. S. Shahparnia and O. M. Ramahi, "Electromagnetic interference (EMI) reduction from printed circuit boards (PCB) using electromagnetic bandgap structures," *IEEE Trans. Electromagnetic Compatibility*, **vol. 46, no. 4**, 580–7, 2004.
281. G Chen, K. Melde, and J. Prince, "The applications of EBG structures in power/ground plane pair SSN suppression," *IEEE 13th Topical Meeting on Electrical Performance of Electronic Packaging*, pp. 207–10, 2004.
282. H. Y. D. Yang and C.-Z. Zhou, "The reduction of electromagnetic interference in RF integrated circuits through the use of metallized substrates," *IEEE APS Int. Symp. Dig.*, pp. 65–8, July 2006.
283. J. Qin and O. M. Ramahi, "Power plane with planar electromagnetic bandgap structures for EMI reduction in high speed circuits," *IEEE APS Int. Symp. Dig.*, pp. 365–8, 9–14 July 2006.

Index

90° reflection phase, optimization for, 113–17

absorbing boundary conditions, 18–22
ADI-FDTD method, 18
angle-updated method, 26–7
Ansoft Designer program, 44, 104–5
Ansoft HFSS program, 44, 63
antenna arrays, 9, 138–48, 243
Archimedean spiral antennas, 243
artificial magnetic conductors (AMC), 70, 85
   metamaterials, 6
artificial neural network (ANN) theory, 240
auto-regressive moving average (ARMA) estimator, 17, 45, 52, 55–6
axial ratio
   circularly polarized patch antennas, 135
   curl antennas, 176–7, 179–80
   dipole antennas, 183

back radiation, 129, 138
bandwidth
   definitions, 158
   for hard operations, 79
   and patch width, 89
   patch antennas, 131–2
   for soft operations, 78
base station antennas, 243
beam scanning, 185–6
beam scanning angle, 242
beam steering, 242
beam switch
   one-dimensional, 188–9
   two-dimensional, 191
beamwidth, single patch antennas, 132
bent monopole antennas, 186–8
bi-anisotropic media, 241
biasing circuits, 189
binary particle swarm optimization, 113
boundary conditions *see* absorbing boundary conditions; Floquet periodic boundary conditions; perfectly matched layers (PML) boundary condition technique; periodic boundary conditions (PCB)

bow-tie antennas, 200, 243
Brillouin zone, 31–3, 68–9
broadband wire antennas, 243

cavity back antennas, 132, 145
characteristic parameters, antenna design, 23
choke rings, 149
circularly polarized antennas, 242
   curl antennas, 171–6, 185
   dipole antennas, 181–2
   patch antennas, 132–5, 153
complementary geometries, 241
constant k_x method, for scattering analysis, 26–30, 45, 71
"conventional" case, 64–5
convergence curves, for miniaturized EBGs, 116–17
convoluted elements, 241
corrugated surfaces, 6, 52–3, 80–1
coupling, mutual *see* mutual coupling, antenna arrays
Courant-Friedrich-Lewy stability, 17, 51–2
cross polarization, 103, 105–7, 189
crosspatches
   center-fed, 228–9
   current densities, 228
   dimensions, effect of, 228
   geometry of, 228
   and surface wave antennas, 226–36
curl antennas, 10, 171–85, 200, 242–3
cylindrical configuration, 241

diamond-shape dipole antennas, 200
dielectric constant, 90, 127–8
dielectric rods, 37, 84
dielectric slabs
   eigen-frequencies, 33, 56
   hybrid FDTD/ARMA analysis, 51–2
   impedance, 47–9
   PCB analysis, 33, 54
   reflection, 41
   scattering analysis, 27
diodes, 124, 185, 189

dipole antennas
  and ground planes, 156–8, 191–200, 209–17
  half-wavelength, 182–5, 193
  infinitesimal, 63
  literature, 242
  project work, 236
  radiation field, 163, 181–2
  relative position of, 161, 191–5
  reflection phase, 161–3
  return loss, 157–8
dipole arrays, 103
dipole frequency selective surfaces, 41–5, 54, 56
direction finding, 244
discretization, 114–15
dispersion curves, 30, 34–6, 206
dispersion diagram, 67–8
  dielectric slab, 204
  mushroom-like EBG structures, 68–9, 92, 204
  uni-planar EBG surfaces, 92
distributed excitation sources, 22–3
double negative metamaterials, 4, 6, 37, 241
double spiral EBG structures, 105
double-arm generic microstrip designs, 241
dual band EBG surfaces, 120, 241
  see also multi-layer stacked EBG designs
dual band surface wave antennas, 223–6, 228–36
dual band wire-EBG antennas, 169–71
dual layer EBG designs, 107–11
dual polarized spare arrays, 243
dumbbell EBGs, 147, 241

EBG resonator antennas, 243
edge position of dipole, 191–5
eigen-frequencies
  from ARMA estimator, 17–18, 51
  for dielectric slab waveguides, 33
  of surface waves, 30
  for waveguides, 23, 25
  wavenumbers, 69
electromagnetic band gap (EBG) structures, 2
  analysis methods, 6–7
  applications, 8–11, 242–4
  cell size, 103
  and curl antennas, 172–80
  history, 239
  manufacture, 130, 241
  performance of, 156–8
  terminology for, 85
elliptical configuration, 241
E-plane coupling, 139–42
E-shaped patch antennas, 127
excitation
  for antennas, 17
  in ARMA method, 49
  in FDTD simulations, 22–3
  in plane wave scattering analysis, 39–40

far field radar cross section (RCS), 240
ferrite substrates, 241
finite difference time domain (FDTD) method
  applications, 14
  EBG analysis model, 7
  excitation, 22–3
  overview, 14
  patch antennas, 128
  representation of frequency gap, 63–5
FDTD/PBC technique, 87
finite element method (FEM), 7, 44, 240
fitness functions, 115–17
Floquet periodic boundary condition, 7, 24
Floquet space harmonics, 67–8, 75–6, 206–9
Floquet theorem, 31, 38
four-arm spiral EBGs, 105–7
fractal antennas, 243
free space dipole position, 193–4
frequency band gap
  FDTD representation, 63–5
  in mushroom-EBG structures, 92
  near fields, 65
  in surface wave propagation, 67–9
  in uni-planar EBG surfaces, 92
frequency domain methods see finite element method; method of moments (MoM)
frequency response, 51
frequency selective surfaces (FSS), 37, 41–5, 240–1
frequency spectrum, dielectric slab waveguide, 33
front to back ratio, 135
full wave simulations, 7

gap width, mushroom-EBG structure, 87, 89
generic algorithm (GA), 120, 242
GPS antennas, 8, 149
ground plane size, curl antennas, 176–7
grounded dielectric slab, 47–9, 51–2, 55, 85
  with periodic patches, 33–7, 203–4
gyro-tropic EBGs, 241

half-wavelength dipole antenna, on PEC ground plane, 193
handset antennas, 243
hard metamaterials and surfaces, 6, 74–5, 240–1
hard operations, 78–9, 81
harmonic control, 242
height of dipole, 195
helical antennas, 200
high gain antennas, and EBG structures, 10–11
high impedance surfaces (HIS), 6, 85
Hilbert curves, 103, 120, 241
holographic antennas, 11
horn antennas, 75, 243
H-plane coupling, 139–42
hybrid FDTD/ARMA method, 45–53

impedance *see* input impedance; surface impedance
implantable biotelemetry systems, 243
incident waves, 2–4, 6, 18
inferior dipole position, 195
inhibition, of propagation, 84
in-phase reflection, 240
input impedance, 23, 174, 219–20
interference reduction, 244
inverted F antennas, 243

Jerusalem cross frequency selective surface, 63, 103
journals, special issues, 238–9

k_x-frequency plane, reflection coefficient, 26–7

leaky mode, 241
left/right handed structures, 4, 241
length of dipole, 195, 210, 212
loop antennas, 242
low profile antennas, 118, 156, 164, 242–3
low profile arrays, 243
low temperature co-fired ceramic (LTCC) technique, 241
lumped element model, 6–7, 240
  for mushroom-like EBG structures, 59–61, 89, 90
  for uni-planar EBG surface, 92

magneto metamaterials, 6
Maxwell's equations, 15
meander-line polarizer, 180
mechanical tuning, 185
metamaterials, 4–6, 37, 239
method of moments (MoM), 7, 17–49, 240
micro-electro-mechanical (MEMS) actuators, 185, 189
micromachining, 130
microstrip antenna, 8, 130, 165–6, 242
microwave circuits, 11, 244
miniaturized EBGs, 117–18, 241
mini-UAV, 244
monolithic microwave integrated circuits (MMIC), 127
monopole antennas, 186–9, 217, 242
multi-band EBG designs, 120, 241
multi-layer EBG designs, 107–9, 241
  see also dual layer EBG designs
multi-layer metallic tripod array, 2
multi-period cell design, 241
multiple unit cell method, 26
mushroom-like EBG structures, 6, 59–61, 240
  classification, 84
  dispersion diagram, 92
  frequency band gap, 92
  and hard/soft operations, 82–4

parameters, 87–90
  as perfect magnetic conductor, 82
  with periodic patches, 203–4
  reflection phases, 92–3
  resonant frequencies, 92–3
  and uni-planar EBG surface, 91–5
mutual coupling, antenna arrays, 138–48, 169, 242–3

near fields
  E-plane coupling, 141–2
  H-plane coupling, 142
  patch fed surface wave antennas, 213–15, 220–2
  source excitation, 240
  superior antenna position, 136
negative impedance loading, 241
negative refractive index metamaterials, 6
network theory, 240
normal incidence, 25, 70–1

oblique incidence reflection, 25, 71–4
observation plane, 71
one-dimensional beam switch, 188–9
open-sleeve antennas, 243
operational bandwidth, 158, 161–3, 228
optimization *see* particle swarm optimization (PSO)

Pade approximation, 49
papers, publication rate, 238
parasitic patches and strips, 127, 189
particle swarm optimization (PSO), 112–19, 124, 242
patch antenna arrays, 138–48
patch antennas, 127–53
  with EBG surround, 130–2
  gain enhancement, 130–8
  with high dielectric substrates, 127–9
  size, 127
  two-layer EBG, 136–7
  and wire-EBG antennas, 166–9
  see also single patch antennas
patch fed surface wave antennas, 217–23
patch units, sizes, 87–9, 95, 110–11
Peano space-filling curve, 120
pencil function, 49
penetration radar, 244
perfect electric conductors (PEC)
  and curl antennas, 172–4
  and dipole antennas, 156–8
  and EBGs, 9, 118
  polarization, 99
  reflection coefficient, 69–70
perfect magnetic conductors (PMC), 70, 79–80, 156–7

perfectly matched layers (PML) boundary condition technique, 18–22
  characteristic parameters, 23
  non-uniform, 20
  numerical errors, 20
  one-dimensional, 18–20
  three-dimensional, 20–2
periodic boundary conditions (PCB), 24–30, 54
periodic composite transmission line structure, 6
periodic ground plane, 75–7
periodic patch loaded slab, 85
periodic transmission line method, 7, 61
  see also transmission line models
photonic band gap (PBG), 240
planar EBG surfaces, 2–4, 84
planar reflectarray antennas, 37
plane wave expansion method, 240
plane wave scattering analysis, 37–45
plane waves
  excitation, 39–40
  impedance, 17–49
  region of, 46
  transmission line modelling, 62–3
polarization
  reconfigurable, 185
  rectangular patch EBG surfaces, 101–3
  in single spiral EBG structures, 104–5
  see also circularly polarized antennas; cross polarization; reconfigurable polarization
polarization converter, 241
polarization-dependent EBG (PDEBG) surfaces, 95, 124
  ground plane with dipole antenna, 180, 182–5
  reflection phases, 241
  reflector, 99
Prony technique, 17, 49
propagation, inhibition of, 84
propagation constant (k_x), 24
propagation loss, 22

quadratic phase criterion, 161, 163
quasi-TEM antennas, 243

radar cross section, 38
radiation boundary condition, 18
radiation efficiency, 9–10
radiation mode, 241
radiation pattern control, 242
radiation patterns, 23
  circularly polarized patch antennas, 135
  dipole antennas, 181–2, 185, 197–200, 211–12
  dual band surface wave antennas, 230–6
  near fields, 136
  one-dimensional beam switches, 189
  patch antennas, 129, 131–2, 167–9

reconfigurability, 185
  surface wave antennas, 212–17, 222–3, 227–8
  two-dimensional beam switch, 191
  vertical monopole antennas, 217
  wire-EBG antennas, 167–9
radio frequency identification, 243
real-number particle swarm optimization, 113
reconfigurable antennas, 185–91
reconfigurable EBG surfaces, 241
reconfigurable polarization, 185
rectangular patch EBG surfaces, 95, 101–3
reflection coefficient, 19, 69
  dipole frequency selective surface, 43–4
  dual band surface wave antenna, 230
  k_x-frequency plane, 26–7
  periodic ground plane, 76–7
  scattering analysis, 23
  surface wave antennas, 228
reflection phases, 69
  dielectric slabs, 204
  dipole antennas, 162–3
  double-spiral EBG structures, 105
  as EBG classifier, 84
  four arm spiral EBG structures, 105–7
  multi-layer EBG designs, 107–9
  mushroom-EBG structures, 71, 92–3, 204
  and patch widths, 88–9
  rectangular patch EBG surfaces, 95
  slot loaded EBG surfaces, 97
  uni-planar EBG surfaces, 92–3
reflector, PDEBG, 99
resonance, dielectric slab, 49
resonance frequencies
  dual-band EBG surface designs, 120
  multi-layer EBG designs, 108–11
  mushroom-EBG structures, 92–3
  and patch width, 88–9
  spiral EBG structures, 103–5, 107
  uni-planar EBG surfaces, 92–3
resonant antennas, 127
resonant cavities, 167, 240
resonant circuit loading, 241
resonant circuit models, 59–63
return loss
  curl antennas, 179
  dipole antennas, 157–8, 183, 195, 197, 211
  microstrip antennas, 143–5
  one-dimensional beam switch, 189
  patch antennas, 128–9, 131, 167
  surface wave antennas, 217, 219–20, 223, 227–30
  vertical monopole antennas, 217
  wire-EBG antennas, 167

S-curve, 116–17
scan blindness, 151, 242
scattering analysis
  characteristic parameters, 23
  constant k_x method, 26–30
  excitation, 17, 22–3
  PCB method, 25
  plane wave, 37–45
  propagation constant, 24–5
scattering angles, 38
semi-EBG ground plane, with dipole antenna, 191–200
series-resonant grid method, 63
sine-cosine method, 25–7
single bent monopoles, 189
single patch antennas
  bandwidth, 131–2
  beamwidth, 132
  gain enhancement, 130–8
  radiation patterns, 131–2
  return loss, 131
  see also patch antennas
single spiral EBG structures, 103–5
size of ground plane, 215–16
sleeve dipole antennas, 200
slot antennas, 103, 242–3
slot loaded EBG surfaces, 97
soft metamaterials and surfaces, 6, 74–5, 240–1
soft operations, 77
space harmonic waves *see* Floquet space harmonics
space-filling curve EBG structures, 120
specific modes, 67
spectral domain analysis method, 240
spectral finite-difference time domain (S-FDTD), 240
spiral antenna arrays, 243
spiral antennas, 200, 242
spiral EBG structures, 103–7, 241
spiral shape frequency selective surfaces, 63
split field FDTD method, 26, 44–5, 71
square lattice patch EBG antennas, 107, 138
stability, Yee equations, 17–18, 51
substrates
  ferrite, 241
  high dielectric constant, 127–8
  permittivity, 87, 90
  removal, 145
  step-like, 130–2
  thickness, 87, 89–90, 128–9
superior dipole position, 193
surface impedance, 46–9
  corrugated soft/hard surface, 52–3, 56
  dielectric slabs, 17, 51–2, 55
  periodic ground planes, 76–7

planar EBG surfaces, 3
surface mode, 241
surface wave antennas
  dipole fed, 209–17
  patch fed, 217–23
  project work, 236
  and vertical monopole antennas, 217
surface wave frequency band gap, 68–9, 158
surface waves
  on dielectric slabs
  modes, 69, 79, 81
  on patch antennas, 129
  region of, 46
  supression of, 8–9, 65, 240
  transmission line modeling, 61–2
suspended patch antennas, 242
swarm behavior, 112–13
system identification (SI) method, 49

tensor-forms, 20–2
textile electronics, 149–51
time domain signal, 23
time domain technique, 240
total-field/scattered-field technique, 39–40
transmission coefficient, scattering analysis, 23
transmission line models, 61–3, 240
  see also periodic transmission line method
transmission lines, 84
transverse wavenumbers, 69
tunable EBG surfaces, 120–4, 241
two-dimensional beam switch, 191
two-layer EBG, patch antenna, 136–7
two-wire EBG antennas

ultra-thin absorbers, 242
unified analysis method *see* hybrid FDTD/ARMA method
uni-planar EBG surfaces, 240
  classification, 84
  compact form (UCEBG), 95, 118, 147–8
  dispersion diagram, 92
  frequency band gap, 92
  and mushroom-EBG structures, 91–5
  periodic transmission line analysis, 62
  reflection phases, 92–3
  resonant frequencies, 92–3
unit cell shifting method, 26
updating equations, 16–18, 51
U-slot microstrip antennas, 127

varactor diodes, 124, 185
vertical monopole antennas, 217
vias, 95, 97–9, 241
volumetric structures, 84

waveguides, 244
　analysis, 30–7
　dielectric slab, 33
　eigen-frequencies, 23
　PCB design method, 25
　propagation constant, 24
　surface waves, 206–7
wavenumbers, 67, 69
wearable electronics, 149–51, 243

wideband EBGs, 241
wideband split field method, 27
wire antennas, 9–10, 243
wire medium, 62
wire-EBG antennas, 166–9
woodpile structures, 2, 243

Yagi antennas, 169, 200
Yee cells, 15–18

Printed in the United States
By Bookmasters